THE bee-friendly GARDEN

Design an abundant, flower-filled yard that nurtures bees and supports biodiversity

THE bee-friendly GARDEN

Kate Frey and Gretchen LeBuhn

photography by Leslie Lindell

TEN SPEED PRESS
Berkeley

contents

Previous spread: Mint in bloom.

Facing page: Native bee on Tradescanta viginiana.

preface

As a professional garden designer, I used to judge the beauty and interest of a garden by the composition of colors, the texture of foliage, and the plants complementary or contrasting forms. I assessed scenes of nature by their ability to evoke awe or by the perfectly arranged composition of plant species. It wasn't until my five-year-old son and I observed the metamorphosis of four brilliantly striped monarch caterpillars on a patch of silvery Indian milkweed in our garden that my concept of beauty and interest changed. Besides the caterpillars, I suddenly began to notice the many species of bees that visited the milkweeds and plants surrounding them: fat and fluffy bumblebees with a variety of stripes, hoods, capes, and bottoms; small bees carrying pollen on the underside of their abdomen; shiny black carpenter bees; stout, long-horned bees; and the most exotic looking: the iridescent green sweat bees. There seemed no end of the flower visitors attracted to just one small area of plants.

Soon I was spending more time watching what was visiting the flowers in my gardens than looking at the flowers themselves. The garden became a place of intrigue. Question after question was generated by each scene: Why were some flowers mobbed by visitors like flies, beetles, and lacewings, others by native bees and honeybees, and still others by just butterflies? What was it about the flower structure, pollen, or nectar that appealed to a specific set of organisms and not to others?

We put up bee-nesting blocks next to the garden and they were filled the first year, each hole neatly plugged with mud or chewed plant

Facing page: Kate Frey's home garden.

stems. Suddenly, the life that was visiting the flowers became an integral part of the beauty and vitality contained in the garden.

From this first milkweed plant, a world has opened and continues to open—endlessly. I have created habitat and bee gardens across California, at the Chelsea Flower Show in London, and in Japan and Saudi Arabia. Each is flower filled, suffused with color and shifting blooms, and is uplifting and engaging to the human visitor. Under the gray skies of London, I planted western native wildflowers and agricultural clovers in orange, deep blue, yellow, red, and white under and around an array of grapevines in a series of three gardens showcasing organic viticulture. Two of the gardens won gold medals and were visited by Queen Elizabeth, demonstrating that organic agriculture and the vibrant colors and soft forms of wildflowers have wide appeal. In Japan, my husband and I created a habitat garden focused on butterflies and bees, composed of a meadow with pastel-colored flowers under the soft, spring leaves of Japanese maples. In the harsh light and dusty heat in Saudi Arabia, on an organic vegetable and fruit farm owned by one of the princes, we planted bright yellow sunflowers, orange and white cosmos, white alyssum, basils, cilantro, za'atar, mint, and fennel to attract and support wildlife. Watching wild bees covered in sticky pollen became our entertainment.

Spending time in a bee garden allows us to step into another world, transcending the everyday routine and entering a place of beauty and anticipation. With these gardens, we develop and maintain a connection to something larger than ourselves—we get to see and know the intrinsic value of the flowers and the lives of the bees that visit them in each season.

—Kate

I come from a long line of gardeners. Some of my earliest memories are of harvesting and canning vegetables with my mother, aunts, and grandmother. I remember vividly trying to decide what should go in the garden that my mother gave me for my own when I was about eight. My Aunt Kitsie, an artist with plants, was responsible for teaching me the importance of design. Her gardens in coastal Massachusetts are inviting, beautiful, and host a variety of local animals.

In the late 1980s, I left a career as an investment banker in New York because I had fallen in love with ecology. A book by James Gleick on chaos theory and work by Dr. Gary Polis, one of my mentors at Vanderbilt, on scorpions on islands in the Gulf of California, in Mexico hooked me. I was so thrilled that all these interesting things were being studied, and I desperately wanted to know more about them. I decided to take a year and see how deep my love of ecology was. Initially, I intended to work on marine systems. However, in ecology, we studied the mutualisms between plants and insects. In particular, I remember my professor James Thomson talking about how different bees matched different flowers. As a lifelong gardener, I could not believe that I had missed this important evolutionary relationship—and I actually didn't even know there was more than one species of bee. That one lecture on mutualisms changed my life. I've been fascinated by the relationships between bees and plants ever since.

I hope you will take the time to become a close observer. Take a moment to learn a few bees, expand your garden, or rearrange it to host more species. I believe that these small actions taken by individuals can effect significant change for landscapes. It just takes a few moments.

—Gretchen

INTRODUCTION

the benefits of a bee-friendly garden

Bee gardens make people happy! Whether you enjoy a brilliant chorus of saturated color, a tranquil sanctuary from the busy world, or a hardworking edible garden, there is a glorious, flower-filled bee garden waiting for you.

Bee gardens are organic and sustainable. They develop healthy and fertile soil, attract beneficial insects for pest control, and support biodiversity, including butterflies and birds. If you are an edible gardener or a seed saver, bees will increase your harvest and generate larger, better fruits and vegetables.

Most importantly, bees are a critical link in the global food chain. Bees are the world's most prolific pollinators. They facilitate the fertilization of the flowering plant on whose pollen and nectar they feed. Without bees, many flowering plants would not be able to reproduce at a sufficiently high rate. Over 70 percent of the world's plants depend on the pollination services of bees, including many nuts, fruits, tomatoes, peppers, or berries. Even plants that we enjoy for their roots, like carrots and beets, rely on bees to pollinate the plants to enable the production of seeds for the next year's crop.

Colony collapse disorder, the phenomenon which has caused a significant and well-publicized decline in honeybee population, has many people alarmed and concerned. What we plant in our home gardens can help by providing more resources to support these bees. Research has shown that cities with more gardens have greater numbers and diversity of bees, and that bees can thrive in urban and suburban environments equally as well as in rural locations. We as individuals can make a difference and enjoy the pleasures of gardening at the same time.

An effective bee-friendly garden doesn't have to be large, nor does it have to be complicated. You can begin with a bare site and develop a bee-friendly garden from scratch, or you can simply incorporate some bee-friendly plants into an existing garden. The plants are easy-to-grow, common garden plants that include natives. You may already have some plants in your garden that are good bee plants.

Despite the fact that a large number of landscape plants have the ability to support wildlife, surprisingly, most American neighborhoods are dominated by a formulaic prescription of lawn, clipped shrubs, pollarded trees, and a small patch of annuals. If you analyze these landscapes for bee-friendliness or attractiveness to any organism besides the ever-hungry lawnmower or hedge trimmer, you will not discover much value. Biodiversity is lacking. This sad situation can be remedied easily. In addition to being full of life, a color-filled bee garden enhances your home in a way no lawn can.

The Beauty of Bee Gardens

Besides providing environmental benefits, the plants and flowers that support bees support humans in crucial ways. We all know flowers make us happy; recently, scientific studies have proven that we respond positively—both physically and emotionally—to plants and flowers. They make us feel more alive because plants and flowers connect us to nature. People recovering from surgery in hospitals who have views of plants outside their windows recover faster and take less pain medication. Spending time among plants, nature, and forests

Oliver —

Gift - from

Bee Friendly

grant.

______Coast ______Horizon

CKOs ______

Last CKO date ______

notes:

lowers levels of cortisol, the stress hormone. The longer we spend in nature, the longer the effects last.

Spending time in a bee garden can be a source of continuous pleasure and nature therapy in your own backyard; observing the colors, shapes, and behaviors of bees is endlessly fascinating. Time slows and the small things catch your attention. Scrutinizing flowers reveals amazing and intricate beauty. Subtle colorations, veining, stripes, dots, and freckles delight people, but also act as lures and guides to attract and guide bees with the promise of nectar, using the opportunity to disburse pollen.

Many people respond to bright color in gardens, while others prefer a paler color palette, or the cool tranquility generated by green foliage and chartreuse flowers. In any given region of the country, there are a variety of bee-friendly plants that can be combined to suit every taste and style. Many of our common landscape plants that support pollinators are easy to grow and attractive. In fact, it is easy to compose a garden entirely of plants—annuals, perennials, ground covers, shrubs, and trees—that bees love.

The important message is to think in terms of profusion. Bee gardens are designed to have a large variety of flowers blooming at one time over a long season. Just as we need to eat a varied diet for health, bees need a variety of pollen and nectar as well. In addition, bees are active as soon as the weather warms and until freezing temperatures cause flowering to cease. They need pollen and nectar during their entire life cycle, so bee gardens are designed to have continuously blooming flowers from early in the season to as late as possible. Compared to most gardens that put on a show for a very short period of time, long-blooming bee gardens are a delight to behold for both people and bees.

Bee Gardens Are Healthy

Both flower and vegetable gardens need predatory insects to keep pest insects under control, and many of the same plants that support bees also support beneficial or predatory insects like lacewing larvae and ladybugs, which eat pests like aphids, thrips, scale insects, and other small pest insects. Plants in the Apiaceae (Umbelliferae) family—plants that produce flowers in umbels, such as parsley, celery, fennel, carrots, and dill—attract and support beneficial insects, and most also support bees. Common garden plants like goldenrod (*Solidago*), milkweeds (*Asclepias*), coneflower (*Echinacea*), black-eyed susans, sunflowers (*Helianthus*), asters, wild buckwheat (*Eriogonum*), mountain mint (*Pycnanthemum*) and shrubs like wild lilac (*Ceanothus*), mock orange (*Philadelphus*), *Prunus* spp., oceanspray (*Holodiscus*), willows, and many native buckthorns (*Rhamnus*) attract a large variety of predatory insects, including bigeyed bugs, assassin bugs, soldier beetles, hoverfly larvae, and parasitic wasps.

Sunflowers provide pollen and nectar for bees and seeds for finches and many other seed-eating birds.

Many native plants attract both beneficial insects and bees. Studies have shown that flowering hedgerows of native plants planted around agricultural fields have about 80 percent beneficial insects dwelling in them and 20 percent pest insects; they also foster many species of bees. On the other hand, weedy field margins containing nonnative wild radish, mustard, and mallow (weeds found in many of our gardens) had the opposite ratio: 80 percent pest insects and 20 percent beneficial insects. Additionally, the weeds can be host plants for a number of vegetable plant diseases, and pest insects traveling between the weeds and the crop plant spread disease to the crop.

Using native plants that bees visit is a much better use of space. Many native plants are excellent garden subjects and deserve to be used more to support native insects and birds. They can serve as attractive garden subjects, or they can be used in gardens to displace weeds, generating attractive, healthier gardens more supportive of wildlife.

Bee-friendly gardens can attract and support a great number of other pollinators and organisms beyond beneficial insects, including butterflies, moths, bats, hummingbirds, and other birds. The foliage, seeds, and berries of bee-garden plants act as food for many native residents and migratory animals. Almost all birds feed their young insects; having flowers that attract insects generates vital food for fledglings. Planting a diversity of plants for pollinators also helps to abate heat, generate soil cover, and increase soil fertility.

A bee garden is healthy not just for bees, but for the environment as well. This is because a bee garden is a balanced environment and is managed organically, using the "right plant, right place" approach, without the use of pesticides and herbicides.

Facing page, top: Fennel (Foeniculum vulgare) *attracts many beneficial insects. Its anise-scented leaves are great additions to food.*

Facing page, bottom: Purple coneflower (Echinacea purpurea)*, an attractive native plant in the daisy family, is an important plant for a number of butterflies as well as bees.*

Systemic pesticides such as neonicotinoids (neuroactive insecticides chemically similar to nicotine) are absorbed by the plant at application and taken up in all the plant tissues from roots to flowers. These and other pesticides can persist for months or years in the plant—and in the soil—after a single application. More than 120 different pesticides have been found in wax and honey samples, demonstrating that bees and many, many other organisms come into contact with a plethora of detrimental chemicals in their quest to gather food. Pesticides often affect more than the target pest organism and can travel by air, water, or on soil particles far from the site where they are applied. Waterways around cities often have high levels of pesticides in the mud where many aquatic organisms live that fish depend on for food. When aquatic insects and fish are affected, birds that depend on both for food are also adversely affected. Organic pesticides also can have this effect as well.

The Importance of Bees in Our Gardens

More than 30 percent of our food crops rely on pollination by bees to produce their fruits and seeds, and 70 percent of the foods we eat benefit from some pollination services. Pollination is, quite simply, the process by which pollen is transferred from the anther (male part) to the stigma (female part) of the flower. Globally, eighty-seven leading food crops require animal pollination, including apples, cherries, pears, peaches, plums, almonds, watermelon, squash, cucumbers, berries, alfalfa, citrus, and many others. Our annual vegetable crops must be planted from seed, and it takes the services of many bees to adequately pollinate them. More than 70 percent of plants worldwide, not just crops, benefit from pollination by animals. Bees are essential to pollinate and therefore perpetuate native plants that clothe the earth's landscape.

Yield and seed viability can decline dramatically with no or few pollinators present. Home vegetable gardens and orchards need the pollination services of bees just as much as large-scale vegetable or fruit crops. Many native plants in and around our home gardens are also reliant on pollinators to transfer pollen from flower to flower.

Facing page: Before and after photos from Whole Foods Market, which removed all produce that comes from plants dependent on honeybees and other pollinators to demonstrate the importance of honeybees. Almost 250 products were removed, including some of the most popular items like apples, avocados, melons, and cucumbers.

Your produce choices *with* bees

Your produce choices *without* bees

Bees are the only organisms that purposely gather pollen to provision nests for their young, thus transferring it from flower to flower in the process. For many plants, they are the world's most efficient pollinators, with behaviors and anatomical features, including body hairs, evolved specifically to further successful gathering of pollen and nectar from plants. Other insect pollinators such as beetles, flies, moths, butterflies, and wasps are less efficient as pollinators on a per visit basis because they are chiefly collecting nectar and may or may not come into sufficient contact with anthers and stigma.

This native bee, covered in dew, is busy pollinating much earlier in the morning and in much wetter conditions than a honeybee would.

Native bees, especially, are very efficient pollinators. It takes only 250 orchard mason bees to pollinate an acre of apples, compared to at least 20,000 honeybees. Many native bees are better pollinators because they are out earlier and later in both the season and the day when weather is cold or wet, while honeybees are keeping warm and dry in their hives. It is not uncommon to see bumblebees and carpenter bees foraging in the rain or out at dawn or dusk. Recent research has shown that the presence of native bees causes honeybees to travel more widely and visit more plants, becoming more effective pollinators.

How Bee Gardens Help Bees

Bees, both honeybees and native bees, are under many pressures from nonnative pests, diseases, pesticides, and lack of forage. All of these factors have contributed to declining bee populations. Each year, commercial honeybee apiarists in the United States lose between 30 to 40 percent of their hives to colony collapse disorder (CCD), a disorder in which apparently healthy honeybee workers suddenly desert their hives. Losses are also seen in hobbyists hives.

Research suggests a variety of stress factors may contribute to colony collapse disorder. Chief among the pressures may be the varroa mite, an inadvertently introduced nonnative parasite, first detected in the United States in the 1980s. These mites suck the blood of bees and in doing so impart many diseases as well as weakening and shortening the bees' lifespan. Pesticides are another stressor, including, but not limited to systemic pesticides like neonicotinoids. Bees encounter pesticides outside the hive as they forage in flowers; in the hive, they encounter contaminated pollen or nectar brought back to it. Even a sublethal exposure of pesticides can adversely affect bee foraging and orientation abilities.

Habitat loss, degradation, and fragmentation are also major issues for bees. Many areas where bees used to gather pollen and nectar from flowering plants are now impacted by urban sprawl, development, and intensive agriculture. The aesthetics of "clean" cause many people to apply herbicides or repeatedly mow roadsides, fencerows, and gardens, removing flowering native plants or even weeds that bees depend on. Farms are increasingly intensively managed, with little area left for native plants or even weeds that contain some sustenance for bees. Some crop fields create a moonscape for bees, where before at least some native flowering weeds or plants in crop fields or margins used to survive.

While home gardens can provide habitat for bees, there is still much room for improvement. Many of our gardens tend to be underplanted, with few bee-friendly flowers in a given area. Plantings are often sparse, with less than a quarter of the soil covered by plants. Mulch or lawn may cover most of the ground, preventing ground nesting bees from finding a site to nest. A garden that is primarily lawn cannot be filled with life. Of the garden flowers we use, not every flower has pollen and nectar that bees can access. Native plants, essential for many native organisms, such as moths and butterflies, are all too rare in our gardens. In addition, studies suggest that many native bees do not visit a number of flowering plants from Central and South America, Australia, and South Africa, though honeybees and some generalist native bees visit some of them.

To put this information in a big picture, in a survey of landscape plants in Berkeley, California, one thousand different species and cultivars of plants were counted, then surveyed for measurable visits by pollinators. Of those 1000 plants, only 128 had measureable visits, and only 50 plants (5 percent) were natives. These ratios are similar across the country and reveal that many of our typical landscape plants do not support pollinators nor beneficial insects or birds. If one extrapolates this information across the millions of gardens throughout the nation that fill 60 million acres in our urban areas and 94 million acres in rural residential areas, it is immediately apparent that we can do more to attract and sustain bees around our homes.

How can the urban or suburban homeowner help? Flowers provide crucial floral resources—pollen and nectar—to both native bees and honeybees. Research by Dr. Gordon Frankie at his Urban Bee Lab (www.helpabee.org) at the University of California, Berkeley, has shown that our urban and suburban gardens can easily support a wide diversity and high numbers of bees, both native bees and honeybees, when planted with a variety of bee-friendly plants.

Cities that have a large number of gardeners and gardens have a large diversity and numbers of native bees. The more floral resources are available, the more bees there are. Dr. Frankie likes to say, "Plant the right plants and the bees will come." In addition, a study from the University of Bristol shows that bees can thrive in towns and cities just as well as (or even better than) they do in farms and nature preserves. Urban and suburban habitats can provide a valuable role in bee conservation.

What we plant in our gardens really does matter. It is important to recognize the potential and power of our home gardens. Each one of us can make a difference in the world by planting a bee garden. Providing viable and reliable food sources that are free of pesticides are crucial to bees' ability to sustain themselves. Our gardens can also provide essential nesting space, which is critical to perpetuating native bee species. Across the nation, in cities and suburban areas, our gardens can serve as viable habitats for bees and delight us as well.

About the Book

This book begins by detailing myths and facts about bees. Bees have specific needs such as nesting sites, food, water, and healthy habitats that must be met for their survival. Information on how to create or protect them in our gardens is essential. Common bee characteristics are detailed and illustrated, bringing these garden visitors into focus. The rest of this book details how to design, plant, and grow a bee-friendly garden with healthy soil and plants, whether you are just adding a few bee friendly plants into your garden, or are creating an entire garden from bare ground. The book finishes with ideas on how to participate in generating, identifying, and preserving bee habitat, and how to join in ongoing research by citizen scientists about honeybees and native bees. Our goal is to help you plan and grow a beautiful and healthy garden that supports the fascinating and vital bee. We can, in our own gardens, create a lovely, engaging space both for ourselves and for bees.

CHAPTER 1

our friends, the bees

Bees are in the order Hymenoptera, a large group of insects. This reflects their shared evolutionary history with ants, wasps, hornets, and sawflies. If you imagine ants with wings, you can begin to see how these insects might all be related. Within the Hymenoptera, bees are most closely related to Apoid wasps, which is why many people confuse bees and wasps. The critical difference between wasps and bees is that bees are vegetarian, eating only pollen and nectar, while wasp larvae are omnivores, eating everything, including insects and spiders. Adult yellow jacket wasps, in particular, are common visitors at picnics, attracted by both the meat and sweets. Bees, on the other hand, are not interested in humans or human food and are happy to just go about their business visiting flowers.

Bees do not aggressively chase down human prey! The best approach if there are bees flying nearby in your garden is to be calm and to not touch them. Bees only sting to defend themselves and their nests. Individual bees generally need firm contact with their bodies, such as being squeezed or stepped on, before they will even attempt to sting. Simply brushing up against a bee as it is flying rarely leads to a sting. Panicking, slapping, and swatting when bees are around, which many people do, actually increases your chance of hitting one and being stung. Additionally, many of our native bees have stingers that are too weak to penetrate human skin, rendering them unable to sting humans, and male bees have no stinger at all. Most native bees are solitary nesters and are not defensive around their nests, although honeybees and bumblebees can sometimes be. For the most part, there is nothing to fear from having bees in your garden.

Previous spread: Honeybee on an aster.

Facing page: Orchard mason bee (Osmia lignaria).

KIDS AND BEES

Children can thrive in close association with native bees, because stings from solitary bees are very unlikely. In fact, there's an elementary school in Oregon that has a large population of gentle *Andrena* bees nesting on its grounds, and the kids call them "tickle bees" for the way they feel when they land on their skin. Personally, we have a flowering raspberry, a favorite of bees, growing near the door of our house in central New York. Despite its close proximity, our family has never had a problem with bee stings.

Honeybees—in other words, social bees with nests—are more likely to sting when defending their nests, but even they aren't interested in stinging when they're out foraging. In addition, kids can learn to stay away from bee nests if they're nearby, just as they learn to stay away from the road—a far greater danger. As grandparents, we're less concerned about bee stings and much more concerned about the loss of pollinators necessary to produce food for our grandchildren in the future.

All in all, it's been amazing how easy and enjoyable it is to live with the company of our native bees, even with young children about.

—Janet & John Allen, www.ourhabitatgarden.org

Honeybees versus Native Bees

When most people hear the word "bee," they automatically envision a honeybee. Yet, honeybees are just one of more than twenty thousand species of bees that exist in the world. Here in the United States, there are approximately four thousand native species, and none of those are honeybees!

Honeybees

Arguably the most familiar bee to humans is the western honeybee, *Apis mellifera*. It has spread now to most continents and is the most important bee partner for agriculture. Honeybees are recent arrivals to the Americas. European settlers of Jamestown, Virginia, brought the first honeybees in 1622, not for pollination services but for honey and wax production. (Technically, these weren't the first members of the

When this swarm of honeybees arrived in her neighbor's yard, Janet Allen of central New York contacted a local beekeeper who safely relocated them, a benefit to everyone concerned, including the bees.

honeybee genus [*Apis*] in North America. A honeybee fossil, complete with hairy eyes, was found in Nevada.) The bees were kept in hives but eventually honeybees escaped from managed hives and successfully colonized in the wild.

Wild honeybee hives can be found in the walls of houses, in trees, or in any large cavity. These colonies build honeycomb and produce and store honey just like managed hives. Swarms are escaped or wild bees that have left their original hive and are following their queen as she searches for a new home. While swarming, the queen emits a pheromone that is all but irresistible to the accompanying worker bees, and they will follow her for miles and days.

If you encounter a swarm, just leave it be. The swarm is generally not aggressive because it is focused on finding a new home. You can contact your local beekeeping group and some local beekeeper will gladly catch the swarm and give it a home in an apiary. If the feral swarms are not captured by beekeepers and moved into an artificial hive, the swarm will find its own cavity. Once the queen settles in, the workers will begin to construct honeycomb from wax and to provision that comb with pollen and nectar for the new brood emerging from eggs the queen lays and to provide resources for the bees of the hive.

Honeybees are social bees, which means that there is a differentiation among the bees into castes. In a honeybee hive, you will find a queen, female bees called workers, and male bees called drones.

Honeybees turn nectar into honey by carrying it back to the hive where it is passed to worker bees called "nurse" bees who "digest" the nectar by mixing it with enzymes to break down complex sugars into simple sugars so it is more digestible and less prone to bacterial spoiling. It is spread in combs to allow water to evaporate, then sealed with wax. Honey represents the food storage of the hive. The stores are what allow the colony to survive cold winters and other periods of low resources.

Native Bees

Native bees are beautiful and diverse. They come in all sorts of colors from the expected black and yellow to bright iridescent blues and greens. They range from the size of a pinhead—tiny *Perdita* bees, only 1⁄10 inch long—to the size of your thumb—about the size of the large carpenter bees.

Most native bees are solitary, establishing their own nest, never living in a colony or hive. For example, leafcutter bees carry a bright dusting of pollen on the bottom of their abdomen and nest in hollow plant stems and other cavities. Squash bees can be found sleeping in squash blossoms early in the morning long before honeybees are out of their hives. Bumblebees are large, furry bees that you may have spotted in your garden, but what you may not have known is that there are 250 species of bumblebees in the world and 46 species in North America. Carpenter bees drill tunnels in soft, unpainted wood for nests.

Squash bees not only collect pollen and nectar from squash flowers, but males also spend the night in the blossoms.

BEE LIFE CYCLE

Typical life cycle of a ground nesting bee. Solitary bees spend most of their life in the nest—the time when you see them flying in your garden is a very short part of their lifespan. Depending on the species and location, pollinating adults my emerge earlier or later in the year than illustrated here.

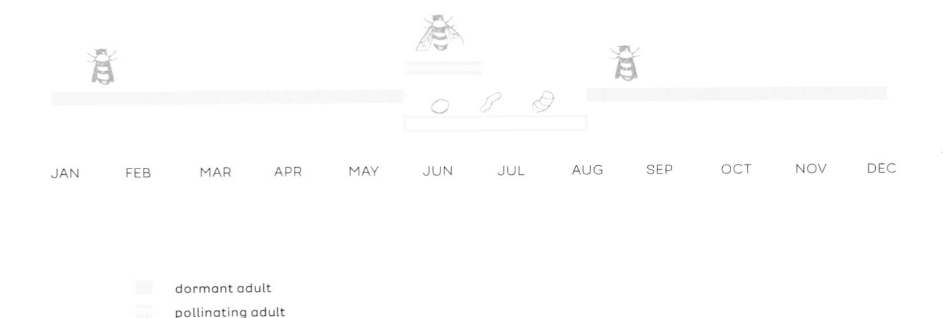

Native bees have very different life cycles from honeybees, which affects their activity in your garden. When we think of bees, we usually envision them busily flying from flower to flower, collecting pollen. In fact, solitary bees spend a lot more of their life in their nests, often underground, than they do flitting around our gardens. How can that be?

All bees go through a series of developmental stages and follow a pattern of metamorphosis, similar to butterflies. They start as an egg that hatches into a larva (similar to a caterpillar), which goes through a pupal stage (wrapped up like a cocoon) before emerging as an adult (a butterfly or bee). These are the periods when bees are in their nests, and for most species, it is the majority of their lifetime.

The solitary bees we see when they emerge in spring or summer as adults were born from eggs laid the previous season or earlier in the same season. Males generally emerge first and females a few days later. This is the beginning of the "flight period," the time that they are active and found in your garden. The flight period of solitary bees, from emergence to the end of their lives, generally lasts only for two to five weeks. They spend the other forty-five to fifty weeks of the year before emergence in their nest.

The emergence of the female is usually tied to the blooming time of the plants she needs to visit to gather pollen and nectar for her offspring. Scientists do not know what causes bees to emerge but think the cues must be related to air temperatures or CO_2 levels in the soil. After emerging, the bees mate and then the female starts to establish her nest (or nests) filled with chambers. This nest can be built underground or in a cavity in wooden post, old tree, or even in the stem of a plant. Once she has started her nest, she gathers pollen and nectar as food for the larvae that will hatch from the eggs she lays in that nest. All bees will take nectar from a variety of plants, but some bees seem to develop better on pollen from specific plants, and most bees show pollen preferences. Bees that gather their pollen only from specific plants are called "specialist bees," while bees that are not as particular and gather pollen from a number of different plants are called "generalist bees."

Once the female has gathered enough pollen and nectar to provide for an egg, she shapes the provisions. In some nests, the female creates a round pollen ball. In other nests, the pollen and nectar are combined into a soupy liquid. In each chamber, she lays a single egg on the provisions of nectar and pollen.

Bee egg on pollen provisions in the cell of a ground-nesting bee.

After two to six weeks, the adults will die but the newly hatched larvae will continue to develop inside their nest. Most of these larvae will enter a state of hibernation that will allow them to survive through the winter without needing additional energy. Bees that follow this pattern are called univoltine, which means that they only produce a single generation per year. Multivoltine species produce more than one generation in single year by having a summer generation or generations that do not enter hibernation and instead develops directly into adults. The second and final generation produced at the end of the summer mimics univoltine bees, entering hibernation for the winter and emerging the following year. Finally, there are bivoltine bees. At the end of the summer, rather than having the adults die off, mated females hibernate through the winter and emerge in the spring to establish the next generation.

While these patterns are interesting, they have important implications for gardens. The critical flight period when bees produce their next generation is when bees need to find flowers. Different species begin their flight period at different points across the year, ranging from earliest spring to late summer. From week to week, there are different bee species emerging and disappearing in your garden, so if flowers are only available for a short period in the spring, there will not be pollen and nectar available for the later emerging species. It also means that practices like tilling the soil in the early spring or removing dead trees can actually kill nesting bees.

Bumblebees are native bees, but they are not solitary and they have a slightly different life cycle. Bumblebees are considered primitively eusocial (meaning they live in a cooperative group in which usually one female and several males are reproductively active and the nonbreeding individuals care for the young or protect and provide for the group). Mated bumblebee queens emerge from hibernating during the winter. These are the large-bodied bumblebees that you see in early spring. She then finds a new nest hole, often an old rodent hole and lays her eggs. For the next few weeks, she works much like a solitary bee, foraging for nectar and pollen and laying eggs. She does all the work on the colony until those first larvae, her daughters, become adults. Then the queen devotes herself to egg laying and the daughters take over the work. The final groups of eggs laid at the end of the summer contain queens for the next year. These queens will mate and forage outside of the nest, putting on enough fat to survive hibernating through the winter. As winter approaches, these queens will dig hibernacula, sites for overwintering, often in a sunny bank and will stay there until the following spring.

Spotting Native Bees in Your Garden

You probably notice honeybees in your garden on a regular basis, but learning to see and identify native bees is a different and rewarding experience. Long-horned bees (*Melissodes*), one of our favorites to watch, are stout and full of character; the male bees have incredibly long antennae, and they have a distinctive, almost horizontal flight as they zip around flowers, landing like a spaceship. And to see an iridescent green sweat bee (*Agapostemon* spp.) emerge from the throat of a brilliant magenta petunia is to see art in action. Once you have learned to identify a few bees, you will find yourself peering into flowers wherever you go to see who is visiting them, and it will change a walk down the block into a much more fascinating journey. Here are a few of the more common native bees. See how many you can spot in (and attract to) your garden!

Yellow-Faced Bees or Masked Bees (family Colletidae, genus *Hylaeus*)

These are generally spring bees. Their bodies are shiny, slender, hairless, and superficially wasplike. They are small, ranging from 0.2 to 0.3 inches long and usually black with bright yellow or white markings on their face and legs. One interesting thing about this genus is that they lack pollen-carrying hairs (scopae) and instead carry both pollen and nectar internally. This way of carrying pollen makes it difficult to assess what flowers they visit (because the pollen cannot be sampled from the bee without dissection), although they are suspected to be primarily generalist foragers.

*Yellow-faced bee (*Hylaeus mesillae*). Note the yellow markings on the face and legs.*

Hylaeus bees nest in stems and twigs, lining their brood cells with a self-secreted cellophane-like material. They lack strong mandibles and other adaptations for digging; thus, many species rely on nest burrows made by other insects.

Hylaeus bees may easily be mistaken for small wasps. To distinguish them from wasps, you can look at the hairs of the bee. All bees have branched hairs that do not reflect light, while wasps have unbranched, light-reflecting hairs that glitter in light. You can only see these hairs under a microscope or a magnifying glass. *Hylaeus* bees are short tongued, but their small body size enables them to access deep flowers.

*Polyester bee (*Colletes hyalinus*). Note the heart-shaped face.*

Polyester Bees or Digger Bees (family Colletidae, genus *Colletes*)

This genus is widespread and common with approximately ninety species of bees in North America. They are found from spring to fall.

Colletes are slender, midsized, hairy bees with pale stripes on their abdomens. Viewed from the front, their heads seems to taper toward the mouth and the eyes are slanted toward each other, making their heads appear heart shaped. They carry pollen in pollen-carrying hairs (scopae) on their hind legs from the upper to the lower part. These bees are primarily specialist bees, visiting a small number of plant species. They are generally found on plants in the aster family, legume or pea family, waterleaf family, borage or forget-me-not-family, mallow family, and caltrope family. *Colletes* are ground nesters, and a few species nest in large aggregations. They line their brood cells with a completely waterproof cellophane-like material, which protects it from fungal attack.

Sweat bee (Halictus tripartitus).

Sweat Bees (family Halictidae, genus *Halictus*)

A widespread and abundant genus, sweat bees are often the most common bees found in bee surveys done in the United States. *Halictus* and the closely related genera *Lasioglossum* and *Agapostemon* are all called "sweat bees" because they are attracted to human sweat and drink it for its salt content. They may land on you for this reason but like all bees, they will not sting you unless you accidentally squish them. But even then, their sting is not strong and is often considered more annoying than painful. *Halictus* are small to medium size, dark brown to black bees. Many species have a dark metallic green sheen when held up to the light. They have bands of hair on their abdomen. Females carry pollen on scopae on their hind legs. Most *Halictus* are generalist foragers, using all sorts of plants from the aster to figwort families. They are very common on composites (daisylike disk and ray flowers) in summer and fall.

Almost all *Halictus* in North America are semisocial ground nesters. Daughters in social colonies remain in the nest and help care for the young. Some species have small nests with a single queen and a few workers, but other species build long-lived nests with multiple queens and hundreds of workers.

Miner Bees (family Andrenidae, genus *Andrena*)

These early spring bees are small to medium size, from 0.3 to 0.7 inches in length. They tend to be moderately hairy, dark, and black or metallic blue or green bees with pale bands of hair on their abdomens. When compared with other groups, their abdomen appears long. Females have large scopae on the upper part of their leg, seemingly in their "armpits." The genus contains both generalist and specialist species. This family has a unique marking: dark lines under the sockets of the antennae that look like eyebrows, as they are often covered in hairs. They are among the earliest bees to emerge in the spring, and they have a very weak sting. Many females can be safely handled. Most *Andrena* are solitary nesters, and they often nest in large aggregations. A few species nest communally, where two or more females share a nest but build and provision their own nest cells. All *Andrena* nest in the ground; they often prefer sandy soil near or under shrubs.

Miner bee (Andrena nigrocaerulea). *Note the blue pollen in the scopa on the hind leg.*

The "eyebrows" of Andrena angusitarsata.

Perdita Bees (family Andrenidae, genus *Perdita*)

A very large genus that is common in summer and fall in western North America, there are approximately 660 species of *Perdita* in North America. These are bees that range in size from 0.1 to 0.4 inches in length. They are usually black, but sometimes metallic green or blue and have abdominal hair bands. They often have yellow or white markings. The profile of their body often seems flat relative to other bees and their wings have fewer veins than other bee wings. These bees are all specialists and are generally solitary ground nesters, rarely communal. Like the miner bees, this group has dark lines under the sockets of the antennae and are among the earliest bees to emerge in the spring. They have a very weak sting.

Male Perdita.

Carder bee (*Anthidium maculosum*) with typical yellow markings on abdomen.

Carder Bees (family Megachilidae, genus *Anthidium*)

Carder bees are widely distributed summer-to-fall bees. These are sturdy, medium-size bees that have beautiful pale markings on their abdomen. Many species in this group specialize on flowering legumes. Like other female bees in the family Megachilidae, the female carder bees carry dry pollen on the underside of the abdomen rather than on their hind legs. They are called carder bees because carding is the process of intermixing wool fibers in preparation for spinning, and these bees use their mandibles to put plant hairs into the cell walls of their nests much like a person preparing wool.

One of the most interesting aspects of carder bee biology is the variation in where they build their cavity nests. They can be found nesting in the old insect burrows or plant stems, or constructing beautiful nests from resin and pebbles. They even have been found in a keyhole.

Leafcutter bee (*Megachile periherta*). Note pollen on the underside of the abdomen.

Leafcutter Bees (family Megachilidae, genus *Megachile*)

Megachile is a large genus that occurs during the summer worldwide with over 130 species in North America. They are commonly known as leafcutter bees because they cut the leaves or flowers of plants and use the pieces to form nest cells. They are the "large leafcutter bees," whereas the genus *Osmia* is the "small leafcutter bees." These are smoky-colored, stout bees with pale stripes of hair on their abdomen. They are medium to large, ranging from 0.4 to 0.8 inches long. *Megachile* are very distinctive. *Megachile* have a stout body shape and hold their abdomens up when on a flower.

Pollen being carried on the abdomen of *Megachile gemula*.

Megachile can be either specialist or generalist foragers. Some species specialize on flowers of plants from the aster family. The alfalfa leafcutter bee (*Megachile rotundata*) is an important pollinator of alfalfa. They are incredibly efficient pollinators because they dive right onto the anthers and stigma and seem to "swim" through the flowers. Since their abdomens are covered in pollen, they easily transfer the pollen from abdomen to the anthers as they "swim" along.

Megachile bees are primarily cavity nesters and nest in a wide variety of habitats and sites. Many species are opportunists; their nests have been found in a variety of human-made structures, and they readily use human-made nests. What *Megachile* are most known for is their leaf-cutting and stem-nesting activities; they all line their brood cells with leaves or petals of plants. In a matter of seconds, a female *Megachile* can cut a leaf and depart for home. Fortunately, this small removal does little to harm the plant.

Evidence of leafcutter bees collecting nesting material from an eastern redbud.

Bumblebees (family Apidae, *genus Bombus*)

Bumblebees are large, furry bees. They are often black and yellow but can have white, red, or orange hairs. Species in this genus have very similar body shapes, but can be distinguished locally by their color patterns. These are social bees. The females store collected pollen in specialized corbiculae or "pollen baskets," smooth, bowl-shaped structures ringed with long hairs on the upper part of their hind legs. Like honeybees, bumblebees have three castes: queens are the largest (0.5 to 1 inch long); workers, which are also female but do not lay eggs, are smaller (0.2 to 0.8 inch long); and males are mid-size (0.3 to 0.9 inch long). There is a subgenus of *Bombus* called *Psithyrus* that are social parasites on other *Bombus* species. A *Psithyrus* female enters the nest of a nonparasitic *Bombus* and kills the queen. The workers of the raided colony then provide for the *Psithyrus* female and her offspring.

A worker bumblebee, Bombus melanopygus, *collecting pollen from a* Ribes *flower.*

Bumblebees fly in cold, rainy weather and are excellent pollinators. They have several physiological adaptations that allow them to fly in bad weather, including the ability to shiver to raise their body temperature. They are often the dominant bee at high elevations. Bumblebee females carry pollen wetted with nectar in their corbicula, so the pollen looks sticky. Bumblebees make a low buzzing sound when flying, usually in a characteristic awkward or "bumbling" pattern.

Bumblebees are among the first bees to emerge in spring and the last to remain active. The large queens are often found foraging on early-blooming species. They are generalist foragers and visit a succession of flowers throughout the flowering season. Many bumblebee species have long tongues that enable them to access nectar from deep flowers such as monkshood, foxgloves, and lousewort. Bumblebees are often used as pollinators of greenhouse tomatoes and cranberries, and colonies are commercially reared just to pollinate these crops. They are also important pollinators of alfalfa, avocados, apples, cherries, blackberries, and blueberries.

In the late twentieth century, biologists began to notice a decline in the abundance and distribution of several bumblebee species, including many bumblebees that were formerly among the most common species in North America. Bumblebee expert Dr. Robbin Thorp (professor emeritus, University of California, Davis) has hypothesized that wild populations of the closely related species *Bombus occidentalis*, *B. affinis*, *B. terricola*, and *B. franklini* were infected by an introduced disease carried by commercially reared colonies of *B. occidentalis* and *B. impatiens*. Four North American *Bombus* species are on the Xerces Society's "Red List of Bees: Native Bees in Decline." Franklin's Bumblebee (*Bombus franklini*) is listed as "critically imperiled" and possibly extinct. The western bumblebee (*Bombus occidentalis*), rusty patch bumblebee (*Bombus affinis*), and the yellow-banded bumblebee (*Bombus terricola*) are listed as "imperiled" and are in sharp decline.

Bumblebees nest socially in annual colonies that they build in abandoned rodent nests, grass tussocks, and other premade cavities. Nests of some species grow to hold more than one thousand individual bees; however, it is more common to have nests that have fewer than thirty bees. Queens fly in late winter or early spring. Workers fly in spring, summer, and fall, and new queens and males fly in the late summer and fall.

Carpenter Bees (family Apidae, genus *Xylocopa*)

Carpenter bee (Xylocopa tabanifomris).

Carpenter bees are common and widespread across the United States, and they are also the largest bees in the United States, often the size of your thumb. They are large bees with round heads and hairy thoraxes. Most species are black, though they may be dark metallic blue or green with blackish wings. Males are sometimes golden yellow. Carpenter bees resemble bumblebees. However, carpenter bees are much shinier and less hairy than bumblebees, and their abdomens, in particular, are shiny in comparison. Females carry dry pollen on scopae on their legs but do not have corbiculae. They have strong and sharp mandibles that help with excavating nests.

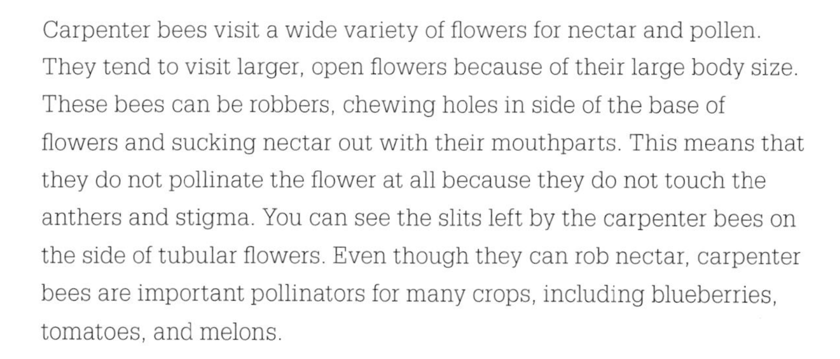

Carpenter bees visit a wide variety of flowers for nectar and pollen. They tend to visit larger, open flowers because of their large body size. These bees can be robbers, chewing holes in side of the base of flowers and sucking nectar out with their mouthparts. This means that they do not pollinate the flower at all because they do not touch the anthers and stigma. You can see the slits left by the carpenter bees on the side of tubular flowers. Even though they can rob nectar, carpenter bees are important pollinators for many crops, including blueberries, tomatoes, and melons.

Most carpenter bees make solitary nests in live or dead wood, or inside the hollow vertical stems of yucca and agave plants. They use their strong mandibles to burrow into pithy stems of wood, often conifers. Some species of carpenter bees have generations that overlap and mothers and daughters may share a nest. Carpenter bees are well-known for chewing nest cavities in decks and the walls of buildings. Male bees are very territorial and patrol their nest or flowering areas vigorously. They will even attempt to chase away a human. Fortunately, male bees do not have stingers, so while you may be surprised; you have nothing to worry about!

What Do Bees Need?

Bees have complex life cycles and specific habitat needs that make them vulnerable to environmental change. Many specialist bees time their emergence to track that of their flowers, so climate shifts that cause bees and their hosts not to be synchronous may really impact these species. Habitat change, pesticides, climate change, and a variety of incompatible land management practices threaten bees. However, at a basic level, bees need three things: a nest site, food, and a healthy habitat.

Nests

In the United States, honeybees are the only species that has a hive with extensive stores of honey. Managed honeybees nest in human-made hives maintained by hobbyists who harvest them for wax and honey and by commercial apiarists who support commercial agriculture. If you are interested in managing your own hive, you should check with your local ordinances to make sure that hives are permitted and get in touch with a local beekeepers group. These groups are wonderful sources of information and often provide beekeeping classes that will help you get started.

A small carpenter bee (Ceratina sp.) exiting her nest, built in a plant stem.

Wild honeybees are really escapees. They find large cavities to settle into that can accommodate a hive of tens of thousands of individuals. While some honeybee keepers believe that wild honeybee populations may develop defenses over time against the varroa mite that could ultimately help managed hives, wild honeybees can have a detrimental effect on native bees and on native plants by preferentially pollinating invasive plant species. Therefore, it's best not to encourage wild honeybees in the United States.

Native bees nest in a range of places and materials: in the ground, in cavities in cement walls, fence posts and trees, or in hollow stems.

There are even bees that nest in snail shells. Most bees are divided into either cavity nesters (which include bees that nest in stems of plants and in preexisting cavities like old beetle holes in dead trees) and ground nesters (aka miners).

Both cavity- and ground-nesting bees need undisturbed areas. These can be a patch of untilled bare ground behind a shed or around plants, areas that are not mowed, or even brush piles. A number of our native species nest in the canes of blackberry, raspberry, sumac, and other similar shrubs, so a fall cleanup that removes those old canes can have a negative impact on those populations. Bees sometimes choose to nest in inconvenient places like in the walls of a house or in an outdoor drawer. Rather than exterminating them, consider practicing peaceful coexistence. Think of it as an opportunity to see and learn about bees up close. For more information on providing bee-nesting sites, including how to build bee nests, see pages 126–30.

Anthophora bomboides *returning home to nest.*

The nest is where a bee will lay her eggs and the larvae will develop. Most native bees are solitary, but there are some solitary bees that build their nests next to each other in large groups of tens or even hundreds of individual nests. For example, every fall, people walking on the bluffs near the beach in San Francisco are very surprised when they suddenly see a lot of good-size bees (*Anthophora bomboides*, a solitary bee that resembles a bumblebee) emerging from the small turrets of sand where they have excavated nests.

Polyester or digger bee (Colletes sp.) *excavating a ground nest.*

MINERS

Bees that dig nest tunnels are called miners. Miner bees usually choose to dig in sunny spots with fine soil that are unlikely to flood. The female digs a long tunnel that can vary from inches to several feet long. Some bees make a central tunnel with cells opening off of it. Other nests have several lateral branches off a main shaft. At the end of the tunnel, the bee creates a chamber that will house her pollen ball and egg. Once the chamber is built, she will gather enough pollen and nectar to provide for her offspring. She often rolls the pollen and nectar into a ball and then lays her egg on the ball. Once the egg is laid, she seals the chamber. Many bees create multiple chambers in the same nest, creating an elaborate network of brood chambers off the central shaft.

A healthy garden for a mining bee will include patches of bare ground. These bees prefer to nest in sandy, loamy soils that do not get saturated with water and are easy to dig. They also have trouble digging through layers of mulch.

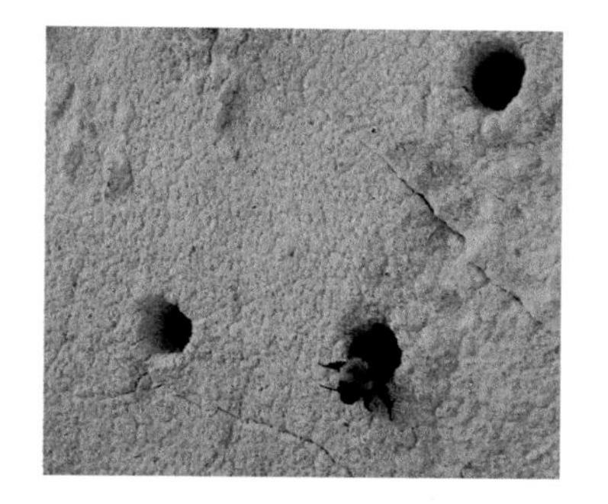

*A cavity-nesting bee (*Anthrophora abrupta*) entering a cavity nest.*

*Leafcutter bee (*Megachile sp.*) returning home with a slice of leaf.*

CAVITY NESTERS

Another group of bees nests in cavities or holes in everything from fence posts to beetle holes to snail shells. Bees that nest in holes usually do not create their own holes. The cavity-nesting bees that borrow a previously dug hole are called secondary cavity nesters because they are the "second" users of that cavity. In these holes, the female bee creates brood chambers filled with a pollen ball just like the mining bees. Brood cells are in a line, filling up the tunnel with the first cell filled being deepest in the cavity. The female bee often puts the female eggs deep in the cavity and the male eggs in the outer part. The female cells take longer to develop so the cells closest to the entry to the cavity contain the males, which usually emerge first. There may be five or six chambers in a cavity.

Leafcutter bees in the family Megachilidae line their nest cavities with neatly cut circles from plant leaves or flower petals. This is the group of bees responsible for those perfectly round holes in rose petals or redbud leaves (see page 28). If you look closely at a fence post, you can sometimes detect a small round hole that has a piece of leaf placed on the end; these are the nests of Megachilid bees. Another group is called mason bees because they use mud to line their cells of their nests. A female mason bee (*Osmia lignaria*), such as an orchard mason bee, will roll a ball of mud and carry it back to her nest in her mandibles. She will then build the dividers for each of her cells using the mud (much like a mason!) and then cap the nest with a mud plug.

*Inside a carpenter bee nest (*Xylocopa *sp.). Note the dividers resembling particleboard between the orange pollen provisions.*

OTHER TYPES OF NESTERS

Two types of bees that do not strictly fit into the miner or cavity-nester categories are carpenter bees and bumblebees. Carpenter bees actively drill nest holes in soft wood. Their nests can be found in garden objects like unpainted arbors, fence posts, some palm trunks, or dead trees and branches. They often nest in the hollow stem of a yucca or an agave. The small carpenter bees in the genus *Ceratina* will use existing small holes in fence posts, old nail holes, or beetle holes for nesting. They plug holes and cap individual cells with chewed sawdust, creating partitions that resemble particleboard after laying an egg on the pollen ball.

Bumblebees nest in a variety of spaces. Some species prefer to build colonies below ground, some above. We've found them in bluebird boxes, underneath stairs, and in a wall filled with insulation. It is more common to find them using old rodent holes in wild areas or tucked in a rock wall. While there are many types of bumblebee nests sold, biologists have yet to come up with a design that reliably attracts bumblebee queens.

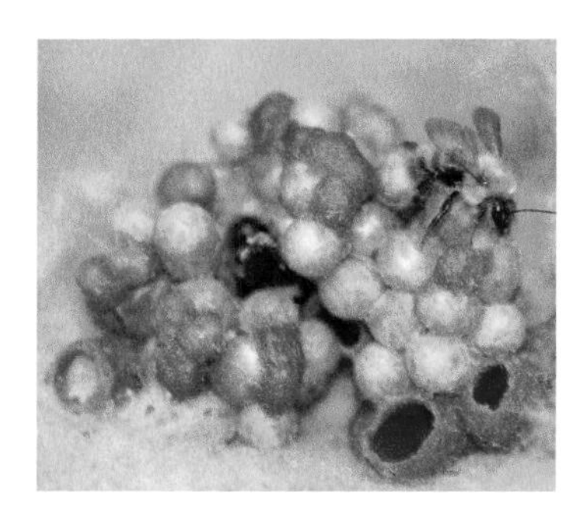

*Bumblebee (*Bombus perplexus*) nest embedded in fiberglass insulation.*

Food

Adult bees use nectar, which is typically 10 to 30 percent sugar, for energy, and they gather pollen to feed their offspring. Different flowers produce nectars and pollens with different levels of nutrition. Pollen is the main source of protein, amino acids, vitamins, and fats for larval bees and is required for their development. Bee-pollinated plants generally have higher levels of proteins in their pollen than wind-pollinated plants.

Bee with pollen on its scopae visiting a gaura.

Nests provisioned with pollen that is protein-rich enables larger offspring. Bees transport pollen in long, stiff hairs (scopae) on their hind legs or under their abdomen, or in pollen baskets (corbiculae) on the hind legs.

Bees go to flowers in gardens to find pollen and nectar, and in the process, they help plants make seeds and fruits by dispersing pollen. Plants, therefore, encourage bees to visit by providing a variety of rewards. Flowers are primarily advertisements that are designed to attract pollinators to pick up or deliver pollen, though sometimes a flower will cheat (false advertising!) and have nothing.

While bees are interested in gathering pollen for their larvae, plants are interested in having their pollen go to the next plant rather than back to the bee nest. Therefore, plants often divide up their pollen in different ways. Some plants only ripen a part of the pollen in a flower each day by maturing a subset of the anthers; some plants have many flowers but only open a few per day. Other plants have elaborate structures that require a large bee to open the flower and get access to the pollen and nectar. Sugars are costly to produce, so plants also tend to ration out nectar. Some flowers refill when drained, others refill on a daily basis. Others produce nectar or make pollen available only in the morning or a specific time of day. There are flowers that trick pollinators by looking promising but offer no reward. In some cases, the plant will have structures that look just like anthers (such as in a male begonia) or mimic a very rewarding flower. Some orchid flowers are shaped like female bees and trick naïve males into actually mating with the flower.

Biologists have identified a variety of plant characteristics that are thought to have evolved in response to pollinators. The marvelous diversity of flower colors, scents, shapes, and sizes can be attributed to the plant's need to transfer pollen.

COLOR

The visual spectrum of bees includes the short wavelength ultraviolet range beyond human vision. This means that a bee's visual acuity is much greater for blues, yellows, and whites than for reds and pinks. Therefore, most bee plants have blue, yellow, or white flowers.

In addition, flowers often have contrasting color marks—lines, patterns, or freckles—that serve as guides to where the bee should go in the flower. These are called nectar guides. Many of these nectar guides are even more apparent under ultraviolet light.

SCENT

Interestingly, there is evidence that suggests that olfactory cues (scent) are important for long-distance attraction, and it is only when bees are up close that they are even able to distinguish flower colors from the background. Bees are attracted to flowers with specific scents. While many moth-pollinated plants (which bloom at night when moths are active) smell very sweet, bee plants tend to be aromatic but less "sweet." Scent is also important for young bees just learning to forage. These young bees tend to rely on scent, but as they get more experience, they shift to relying more on visual cues.

SHAPE AND SIZE

Flowers come in a variety of shapes and sizes. The simplest flowers are flat and open. Many of these produce their nectar in nectaries between the petals. This means that a bee will land on the flat surface and move its body around to get access to the nectaries. The pollen it gets on its body is often an incidental effect of this movement to access the nectary.

Facing page: A group of foxgloves (Digitalis) showing the colorful nectar guides whose purpose is to guide bees to the pollen and nectar contained in each flower.

Another basic shape is an open cup. Open cup–shaped flowers, such as a poppy or buttercup, are also easy for all bees to access. These flowers usually have a large ring of anthers in the center, and you can

*The annual pincushion flower (*Scabiosa atropurpurea*) is composed of many small florets clustered together in a single head, each containing pollen and nectar for bees.*

watch bumblebees almost rolling around in the pollen. These flowers often only offer pollen as a reward, which means that they are primarily visited by bees.

More complicated shapes, like that of a pea or lupine, limit which bees can access the nectar or pollen in that plant. These flowers often require heavier, larger bees like bumblebees for pollination. These large bees are strong enough or heavy enough to push open the flower. For example, sage plants have flowers with lobes that function as a landing platform where a bee can land to enter a flower, often called the lip. The pollen is hidden in the upper part of the flower, called the keel. When the bee pushes its way into the flower to reach the nectar while standing on the lobe, the stigma and stamen pop out of the keel and pick up any pollen already on the bee's back and places their own pollen on the center of the thorax of the bee. When the bee moves to the next flower, the process repeats and pollination happens.

Flowers with a center disk, such as sunflowers, blanket flower (*Gaillardia*), cosmos, single-flowered dahlias, and other members of the daisy family (Asteraceae), are favorites of many pollinators. When you look at a sunflower, you are actually looking at tens or hundreds of flowers. The center "disk" part of the flower is made up of tiny flowers, each with its own nectar and pollen. The outer flowers, which are petals, are called ray flowers and function primarily as attraction. There is a second group of plants in this family that more closely resembles brushes. These are the flowers of plants like goldenrod or joe-pye weed. In this case, there are not disk and ray flowers; instead the small flowers simply make up a compound inflorescence. Having a lot of flowers combined into a single head like the plants in the daisy family and pincushion flowers provides a lot of nectar and pollen resources in a small area. From the plant perspective, within a single inflorescence, individual flowers can open slowly over time, allowing resources to be spread over a number of different visitors and hopefully deposited on a number of different mates.

Another interesting group of flowers are those plants with tubular flowers. Usually, nectar is found at the bottom of these tubes. The length and width of the tube can limit which pollinator species can access the nectar. Many of these flowers are visited by moths and butterflies who have very long tongues. However, bees also can be pollinators of flowers of this shape (see page 88). This flower shape is typical of plants in the mustard family (Brassicaceae), such as wallflowers (*Erysimum* spp.), and the pink family (Caryophyllaceae), such as campions (*Silene* spp.).

Plants like Queen Anne's lace (*Daucus carota*) and sea holly (*Eryngium* spp.) have a slightly different flower shape. If you see a flower that has a number of short flower stalks (called pedicels) which spread from a common point, ending in tiny flowers, it is known as an umbel. This is characteristic of the celery, carrot, or parsley (Apiaceae) family. One mostly finds small bees, like the tiny *Lasioglossum* and *Perdita*, using these flowers.

Some of our favorite types of flowers are in the nightshade (Solanaceae) family. The anthers in these flowers form a central cone. The pollen is only released from the anthers when the flower is vibrated at the right frequency. This is called "buzz pollination." You can actually take a tuning fork for the note of C and touch it to a flower to see the pollen released. This is a way to restrict the pollen from being available to all visitors. It turns out honeybees cannot buzz, so it is primarily bumblebees and bees in the genus *Anthophora* that visit these flowers. Several of the most common garden vegetables are in this group, including tomatoes and peppers (see page 105).

Healthy Habitats

Bees need a space that avoids or minimizes the use of pesticides and herbicides. Exposure to pesticides can affect bees outright by killing them; sublethal doses can cause bees to become disoriented and weak or may affect their ability to fly. Toxins brought back to the nest in nectar or pollen can affect developing larvae, or may impair the queen's egg-laying ability. There are many nonchemical ways to treat your plants, starting with tolerating some insect damage and some weeds. Use the "right plant, right place" approach to choosing which plants to grow in your garden. Plants that are appropriate, or adapted to a region's climate, soils, and pest pressure are more likely to thrive and not require intervention such as pesticides. You can think of weeds as resources for your bees when the flowers in your garden are not doing the whole job. More information on how to eliminate the use of pesticides in your garden can be found on page 130.

Facing page: A variety of bee-friendly flower shapes. From left to right,

Top: Lupine, sunflower.

Center: Mint, Angelica stricta *'Purpurea', evening primrose (Oenothera).*

Bottom: Salvia 'Purple Majesty', Penstemon 'Garnet'.

CHAPTER 2

plants for your bee-friendly garden

A successful bee-friendly garden is determined by selecting the right plants for both your garden and the bees. Bee-friendly plants can be annuals, perennials, biennials, shrubs, or trees. Most bee-friendly plants support a diversity of bee species. There are many bee-friendly plants that are easy to grow and long-blooming, so your garden will be filled with flowers for as long as possible with the least amount of effort—leaving you plenty of time to go out and peer into the flowers to see who is visiting.

Some people are discriminating in terms of color preferences, preferring pastels over bright primary colors, or all white gardens, or specific color combinations. There are bee-friendly plant possibilities to suit every taste. Even those who relish chartreuse will find a range of bee-friendly plants to choose from. Bees see in the ultraviolet end of the color spectrum, and prefer flowers in blue, white, pink, and yellow for this reason, but will also avidly visit orange flowers and some red ones as well.

The only constraint on plant selection is that you need to choose bee-friendly plants that will thrive in your soil and climate. Consult a zone map at http://planthardiness.ars.usda.gov/PHZMWeb, along with the plant lists on pages 190–206 to find plants that work best in your climate.

Previous spread: Sedum telephium 'Autumn Joy'.

Facing page: The very fragrant common milkweed (Asclepias syriaca) *is native to much of the Eastern United States.*

If you look closely, you will see that there are different types of bees on different flowers. Both honeybees and native bees show clear preferences as to what plants they will visit, depending on bee size and shape, flower size, color and pattern, flower structure, and access to and composition of pollen and nectar. Sufficient flower patch size is important too (see page 120). Some plants, such as conifers, oaks, olives, grapes, and grains are wind pollinated and produce low-quality (low-protein content) pollen that bees do not usually gather. With some exceptions, bees prefer to expend their energy more productively on plants that produce high-quality pollen. In addition, all nectars do not appeal to all bees. Bees exhibit an excited behavior when visiting some plants that produce copious amounts of nectar, like the highly fragrant, common milkweed (*Asclepias syriaca*), mountain mint (*Pycnanthemum muticum*), the exotic viper's bugloss (*Echium plantagineum* 'Blue Bedder'), and 'Autumn Joy' sedum (*Sedum telephium* 'Autumn Joy') and while collecting pollen from poppies.

Viper's bugloss (Echium plantagineum 'Blue Bedder') *is an annual that both honeybees and native bees find very attractive. It can reseed in your garden.*

In some cases, we don't know why bees prefer some flowers and not others. In general, while honeybees tend to visit plants from places like Australia and New Zealand, South Africa, and Central and South America, most native bees do not. Native bees often do not as enthusiastically visit plants from the Mediterranean, such as thymes, rosemary, sage, and oreganos, while these plants are absolute honeybee favorites, not surprising considering that honeybees are native to the Mediterranean. Plants native to North America are the best choices for native bees, but many native bees are generalists and will also visit specific nonnative plants.

Sedum telephium *blooms mid to late summer, a time that is important for honeybees to have flowers available.*

Attractive and Easy-to-Grow Bee Plants

Trees, shrubs, perennials, and annual plants are all possibilities for inclusion in a bee garden. People with little time may prefer to use more shrubs and trees in their gardens, while others who wish for a colorful display all summer can chose a diversity of perennials, with some shrubs included. Some people embrace annual flowers and the greater amount of attention they can require in exchange for the sheer

Facing page: Mountain mint (Pycnanthemum muticum) *is a favorite plant for many bees, especially honeybees. Its silvery bracts, profuse white blooms, and upright form make it a garden favorite.*

volume of bloom and profuse color they provide. Some people have conditions where annual wildflowers will establish easily and can reseed. Vegetable gardens can be enhanced with attractive, bee-friendly annuals that grow in the same conditions as vegetables do.

Each plant group needs placement in a garden based on how large they become and basic cultural requirements. For instance, annuals live not for one year as their name suggests, but for one specific season, such as the frost-free period of summer or the cool weather of spring. In mild winter climates, an annual display can be changed twice a year. Many perennials require some maintenance during the course of the growing season and look best when this is attended to. Grouping early bloomers with later bloomers can help ensure a garden with interest and floral rewards all season. The burgeoning foliage and flowers of later bloomers can cover the declining appearance of early bloomers.

Some shrubs require initial training, then are fairly maintenance free, while others need some regular pruning. Some trees can live far longer than our lifetimes and can grow over 100 feet tall; others fit comfortably in a small garden. All of these plant groups obviously require completely different approaches to make a thriving and positive contribution to your garden's function and composition.

Annuals

Annuals are climate-specific plants that generally grow, bloom, and die in one season, usually winter or summer. Few annual plants tolerate year-round weather conditions. Winter annuals or hardy annuals are frost hardy to a certain extent, and are generally best grown in the cool season in hot climates because most don't thrive in hot weather. In cold areas of the country with cool summers, these may be grown in the spring and summer. In zones 8 to 10, they are planted in fall, grow through winter and bloom in the spring as temperatures warm and the days lengthen. Summer annuals, referred to as tender annuals, grow only in frost-free months of the year because they do not tolerate frost. They are generally planted in spring after all danger of frost has passed. The degree of heat and humidity each species tolerates varies.

Annuals' main goal in life is to grow quickly, bloom abundantly, and produce as much seed as possible as quickly as possible, which makes them obliging and colorful garden subjects. Many are very easy to grow. Most bloom best with adequate soil fertility and water, and respond to compost being incorporated in the soil by attaining larger size and blooming longer. Some will reseed in your garden, eliminating the need for replanting.

ANNUALS IN THE GARDEN

Though annuals usually require more soil preparation than other plants, the vibrant color, profuse bloom, and array of flower types make them a must for a garden. Many make good cut or dried flowers. For every region and season, there is a symphony of possibilities—from recumbent to exploding in the air, pastels to brilliant orange, and flower shapes that create a scene in themselves. Few plants are so rewarding. A whole area can be devoted to them, or they can be dotted in a perennial bed for added color and bloom or to extend the flowering season. There are many annuals that bees readily visit, and the sheer number of flowers on each plant makes them very desirable pollen and nectar sources.

Do not select annuals that are dwarfed, excessively doubled, or early blooming as these overly bred varieties are not as robust as more natural plants. Some have lost their attractiveness to bees due to an overabundance of petals preventing access or displacing plant reproductive structures. Others may have been bred for decreased or no nectar or pollen production.

Many of our native annual wildflowers are amazing garden subjects in terms of unique and delicate flower color and form, but are little grown in this country. It is quite surprising to see many wildflowers from the United States grown in Japan in gardens and pots outside people's houses—varieties we do not grow here. Just because wildflowers are not easily available as plants does not mean they are not worthy and easy to grow garden subjects. Many annual wildflowers grow very easily from seed and are readily available through regional seed companies. Some are available from mail order nurseries such as Annie's Annuals and Perennials (www.anniesannuals.com).

It is best to put annuals in an area where it is easy to cultivate the soil each year, as most grow best with loose, fairly fertile soils, plenty of sun, and regular water and weeding. Even the desert wildflowers are more robust and longer blooming when given more soil fertility in the form of compost and water, though the degree of both should be adjusted cautiously. If annuals are bound for a perennial bed, make sure to incorporate compost around the plant when they are planted. This method allows you to have thriving annuals without digging up a whole area. Slugs, snails, earwigs, and pillbugs readily eat annuals, especially young plants. At least some of these pest species often live under or in the shelter perennials provide. Sprinkling Sluggo, a nontoxic slug and snail killer around the young plants and checking regularly for depredation until the plants are larger, should ensure success.

When plant species are adapted to a site, some annuals may reseed, including poppies, phacelias, spider flowers, clarkias, toadflax, honeywort, borage, viper's bugloss, blanket flowers, sunflowers, black-eyed susans, and bachelor buttons, so you may not need to replant each year.

Facing page: Bachelor button (Centaurea cyanus) is a bee-friendly annual that comes in a variety of attractive colors.

GROUPINGS OF ANNUALS

A whole bed of annuals is a joy to behold, especially if exuberance and carefree abundance appeals to you. The effort required to grow annuals is really worth it—in both small and large areas, in terms of the color and sheer volume of bloom generated in spring or summer.

A favorite and easy summer combination of large, summer annuals is purple spider flower (*Cleome*), Mexican sunflowers (*Tithonia rotundifolia*), and orange Klondike cosmos (*Cosmos sulphureus*) for a soft purple and bright orange explosion. A sophisticated combination of orange and white is to combine white cosmos (*Cosmos bipinnatus*), Mexican sunflower, and Klondike cosmos together. Many bee species visit the Mexican sunflower as well as butterflies. Gardens planted with cosmos are good places to observe bee behavior because many bee species visit the open and compound inflorescences, making the bees easy to see.

For a spring show in cool weather, few flowers can rival poppies. All the colors that Shirley poppies (*Papaver rhoes*) come in—white, pink, and red—have delicate, crinkled petals that invite close examination. Look inside the blooms in the morning and watch frenzied bees going around and around the feathery anthers seeking pollen, the primary floral reward produced by poppies. Grown on their own, or incorporated into a perennial planting, scarlet Shirley poppies bring a garden to life where ever they grow. Because the petals are large and thin, they capture light and glow like scarlet lanterns. Breadseed poppies (*Papaver somniferum*) and peony poppies (*P. paeoniflorum*, sometimes refered to as *P. hybridum*) are the ultimate poppies. They are sumptuous in both leaf and flower. The leaves are a wonderful frilly, glaucus gray, and the flowers are each like a Flemish still-life painting. Each detail of petal, color, and profuse, velvety anthers invite contemplation. Bees avidly wallow in the anthers, collecting pollen. Breadseed and

*Facing page: This grouping of heat-loving annuals, including sunflowers, Mexian sunflowers (*Tithonia rotundiflora*), orange Klondike cosmos (*Cosmos sulphureus*), and white cosmos (*Cosmos bipinnatus*), creates a vivid chorus of bee-friendly color over the summer months.*

peony poppies come in flower colors from white to lilac, burgundy, and black. A couple of the most exquisite is a frilly white-and-red confection called *Papaver hybridum* 'Daneborg' and a shaggy ruby-purple variety called *P. hybridum* 'Drama Queen'. They are very easy to grow.

If you enjoy blue flowers, phacelias are arguably the favorite wildflowers for bees in spring (or summer in cool summer areas). Many are native to California, the Southwest, Texas, and Nevada—though other species grow in some areas of the United States. Lacy phacelia (*Phacelia tanacetifolia*) is grown worldwide as bee fodder, and this highly valued species has been grown in Europe since the nineteenth century. It reseeds when conditions suit it.

California poppies bloom for many months and often reseed. When combined with the blue phacelias or viper's bugloss (*Echium plantagineum* 'Blue Bedder'), they create a lively combination of warm and cool colors. A low-growing flower for cool conditions is baby blue eyes (*Nemophila menziesii*). It looks wonderful in a pot or combined with California poppies (*Eschscholzia californica*), honeywort (*Cerinthe major* 'Purpurascens'), Chinese houses (*Collinsia heterophylla*), or any of the smaller phacelias.

Another cool season and obliging annual native to North America but grown in the cooler areas of the UK and Europe since 1834 for bee fodder is meadowfoam (*Limnanthes douglasii*), a plant aptly named poached egg flower because the blooms often have yellow centers and white petal margins. It is a low-growing annual that grows to about 1 foot. It grows naturally in wet areas and is almost completely covered in flowers, hence the name meadowfoam. In some areas, it can fill an entire valley, creating a giant pool of spilled milk when seen from afar. In areas with cool summers, when planted in spring, it may bloom a good part of the summer.

Facing page: The soft blue blooms of lacy phacelia (Phacelia tanacetifolia) *appeal to many bee species and are fragrant. Combined with orange California poppies they create a compelling composition.*

BEE-FRIENDLY ANNUALS

Some common and good examples of bee-friendly annuals are:

Hardy and half-hardy annuals

Bachelor button (*Centaurea cyanus*)

Black-eyed susan (*Rudbeckia hirta*)

Blanket flower (*Gaillardia*)

Borage (*Borago officinalis*)

California poppy (*Eschscholzia californica*)

Honeywort (*Cerinthe major* 'Purpurascens')

Iceland poppy (*Papaver nudicaule*)

Peony poppy (*Papaver paeoniflorum, P. hybridum*)

Phacelia

Pincushion flower (*Scabiosa atropurpurea*)

Prickly poppy (*Argemone*)

Shirley poppy (*Papaver rhoes*)

Toadflax (*Linaria maroccana*)

Viper's bugloss (*Echium plantagineum*)

Tender annuals

Basil (*Ocimum basilicum*)

Cilantro (*Coriandrum sativum*)

Cosmos, both pink and white(*Cosmos bipinnatus*), and orange (*Cosmos sulphureus*)

Cuphea, various

Mexican sunflower (*Tithonia rotundifolia*)

Plains coreopsis (*Coreopsis tinctoria*)

Spider flower (*Cleome hasslerana*)

Sunflower (*Helianthus annuus*)

Facing page, from left to right,

Top: Shirley poppies (Papaver rhoes), *Iceland poppies* (Papaver nudicaule).

Center: Lacy phacelia (Phacelia tanacetifolia), *honeywort* (Cerinthe major '*Purpurascens*'), *blanket flower* (Gaillardia).

Bottom: Plains coreopsis (Coreopsis tinctoria), *spider flower* (Cleome hasslerana).

ANNUAL WILDFLOWER MEADOWS Wildflower seed is often sold with expectations of generating fields of wildflowers from simply broadcasting seed around one's garden or field. The truth is this: specific conditions must be met for successful germination and growth of seeds. Lack of weeds, correct soil type, temperature range, adequate fertility, and regular moisture are all necessary to get a good stand of them. Usually sites must be tilled up and compost incorporated unless the soil is already loose and friable and has fairly good fertility. Sites may be tilled or dug a couple of times to kill germinating weed seedlings before seeds are planted to give wildflowers the upper hand. Regular moisture is always required.

Left: A quiet meadow of white California poppies (Eschscholzia californica) *and meadowfoam* (Limnanthes douglasii).

Right: A brilliant and bee-friendly meadow of Shirley poppies (Papaver rhoes) *and farewell-to-spring* (Clarkia).

Kathy Keatley Garvey's small backyard bee garden.

VERY SMALL BEE GARDENS

Anyone who thinks very small, urban spaces can't support a bee garden should visit Kathy Keatley Garvey, a senior writer (and bee enthusiast) at the University of California, Davis, Entomology and Nematology Department. She is the author of the popular *Bug Squad Blog: Happenings in the Insect World*. Her garden is composed of a series of small patios placed around a Spanish-style house built in 1920. Her bee garden is right outside the kitchen and is surrounded for privacy by cherry laurel hedges where many birds nest. There is a small pond backed by a variety of artfully pruned Japanese maples.

In the middle of the garden space is the bee garden, only about 20 by 25 feet. Catmint, California buckwheat, blanket flower, bulbine, lavender, basil, echium, gaura, salvias, sedum, thyme, oregano, rosemary, and coyote mint all grow around the brilliant orange, summer annual, Mexican sunflower. By midsummer, many of the perennials have finished their peak bloom, but the Mexican sunflowers are just hitting their midsummer stride and are crowned by many brilliant orange blooms. The number of bees visiting them are visible from her kitchen steps, and when you get up close, long-horned bees (*Melissodes*), honeybees, bumblebees, and many others keep the air stirred up in the otherwise small, tranquil space. There is a long-horned bee species (*M. agilis*), a name that aptly describes the rapid darting and zipping of their flight—resembling the action of a miniature Star Wars spaceship—the males in pursuit of females and chasing away the competition. There are also many cape fritillary butterflies, a large, deep orange-and-black species whose caterpillars feed on passion vine (*Passiflora*) planted nearby.

Kathy puts a folding chair in the middle of the garden so she can comfortably sit while she pursues the activities of the garden denizens with her camera—their activities documented daily in her fascinating and titillating *Bug Squad Blog* at http://ucanr.edu/blogs/bugsquad. The antics of the garden visitors rival the news in a major newspaper. Her garden shows just how many bees and associated wildlife can be supported by the flowers in one small garden in one urban backyard.

Perennials

Perennials are plants that live for one or more years. Some perennials are short lived and may perform weakly the second year, while others may live for many years. Most perennials die to the ground in winter and then return in the spring. These are referred to as herbaceous perennials. However, some herbaceous perennials have green leaves all year—such as some penstemons (or beard-tongues). Tender perennials are those plants that will live for more than one year if they are grown in a frost-free climate. They are generally reliably perennial in zones 8 to 10 or 9 to 10. In other areas of the country, tender perennials will not survive the winter, but instead are grown as annuals and replanted in spring. Winter in the perennial garden is apt to be pretty bare but is made up for in summer with exuberant bloom.

Some perennials need dividing every few years because the interior of the plant declines while new growth continues on the perimeter; others spread horizontally. A number form a stalwart clump that lasts for many years. Most flower for a specific portion of the year, such as spring, summer, or fall. A few flower for two to three seasons, making them attractive garden subjects for bees and ourselves. Cutting off the spent flowers (deadheading, see page 81) encourages more blooms.

Many perennials have more than one blooming season. Some salvias (or sages) bloom in spring, take a break midsummer, and bloom again in the fall. Others bloom from midsummer through frost. Penstemons have their main bloom in the spring, then will flower to a lesser extent all summer. Catmints' main bloom is in early summer, followed by a second flush mid- to late summer. Many asters bloom in late summer, except *Aster* × *frikarti* 'Monch', a lovely lavender aster that blooms all summer.

The classic English perennial border, cottage garden, and mixed border of shrubs and perennials popular among garden enthusiasts are usually composed mostly of flowering plants, with an emphasis on perennials, but including some shrubs. Perennials come in such a vast array of sizes, plant and flower forms, colors, and foliage that it is impossible to categorize them in a short space. Some perennials may be 10 feet tall, such as the aptly named giant sunflower (*Helianthus giganteus*), or may be a well-behaved froth about 1½ feet tall like calamint (*Clinopodium nepeta*). It is just this array of possibilities of form and color that make perennials such useful and artistic additions to the garden.

Combine plants with contrasting forms for interest. Perennials with softly mounded forms like hardy geraniums, catmints, coreopsis, lavender, and salvias combine well with upright or vertical perennials like the mulleins, foxgloves, ironweed, joe-pye weed, lupine, cup plant, tall verbena (*Verbena bonariensis*), or the wood salvias (*Salvia nemerosa* or *S. sylvestris*). Spiky plants like the sea hollies (*Eryngium*) generate textural interest. Culver's root (*Veronicastrum virginicum*) has ethereal flower spikes in white to lilac and grows well in moist situations. If you enjoy elegant, silver-white flowers, the mountain mint (*Pycnanthemum muticum*) is stunning for months and is visited by many bees.

Penstemons are an underutilized group of plants and contain species and cultivars suitable for almost every climactic area of the nation. One favorite is the foxglove penstemon (*Penstemon digitalis* 'Husker's Red'), with burgundy, shiny foliage and charming white flowers. Another is the Rocky Mountain penstemon (*P. strictus*) a sturdy study in grape about 2 feet tall. Bergamot (*Monarda fistulosa*), a favorite bumblebee plant, has bright pink-magenta flowers held above the foliage in soft, spidery forms. Other plants with vertical interest are the daisy flowers. Both the pinky-orange shades of the purple coneflowers and the brilliant yellow black-eyed susans create dynamic structures that seem to hover like groups of floral helicopters. Common sneezeweed is a soft and velvety daisy, with flowers from brilliant yellow to deep mahogany.

Above, left: The knotweed Persicaria amplexicaulis *'Taurus' has vivid raspberry blooms in late summer that combine well with perennial sunflowers or deep-colored asters blooming at the same time.*

Above, right: Many cigar plants, especially the cultivar Cuphea × *'Kirstens's Delight', are bee and hummingbird friendly and bloom for months. In cold winter climates, they can be grown as annuals.*

Some fall-blooming plants include black-eyed susans, sneezeweed, perennial sunflowers, and asters, as well as knotweeds, which are excellent and very floriferous late-blooming plants. They come in deep pink to reddish shades and the abundant flowering spikes create a show for weeks. 'Taurus' knotweed (*Persicaria amplexicaulis* 'Taurus') is a striking raspberry-red knotweed that doesn't spread. Cupheas bloom for months and have very attractive flowers of pink, purple, and red. They can be grown as annuals in cold climate. A bright yellow daisy from the Southwest and Mexico grown in many areas of the country as an annual is bidens (*Bidens ferulifolia*). It is perennial in frost-free areas. It blooms nonstop all summer, creating a solid mat of yellow, and attracts many bees, particularly the long-horned bee. Single dahlias are very attractive to honeybees and, in some areas, bumblebees. They come in a variety of colors and some even have very attractive burgundy foliage. They bloom for many months and are very easy to grow, fitting in well in a perennial border, as a summer hedge, or as a feature in themselves.

BEE-FRIENDLY PERENNIALS

Some examples of good perennials that have colorful and interesting flowers highly attractive to bees, plus are easy and obliging garden subjects are:

Anise hyssop (*Agastache foeniculum* and hybrids)

Aster (*Aster*)

Bergamot (*Monarda fistulosa*)

Black-eyed susan (*Rudbeckia*)

Blanket flower (*Gaillardia*)

Butterfly weed (*Asclepias tuberosa*)

Calamint (*Clinopodium nepeta*)

Catmint (*Nepeta*)

Coreopsis (*Coreopsis*)

Culver's root (*Veronicastrum virginicum*)

Cuphea (*Cuphea*)

Dahlia (*Dahlia*)

Germander (*Teucrium*)

Goldenrod (*Solidago*)

Hardy geranium (*Geranium*)

Joe-pye weed (*Eutrochium maculatum* and *E. fistulosum*)

Knotweed (*Persicaria amplexicaulis*)

Lavender (*Lavandula*)

Mountain mint (*Pycnanthemum muticum*)

Oregano (*Origanum*)

Penstemon or beard-tongue (*Penstemon*)

Purple coneflower (*Echinacea purpurea*)

Purple prairie clover (*Dalea purpurea*)

Purple toadflax (*Linaria purpurea*)

Salvia or sage (*Salvia*)

Sneezeweed (*Helenium autumnale* and other species)

Thyme (*Thymus* spp.)

Yellow coneflower (*Ratibida pinnata*)

Facing page, from left to right,

Top: Purple coneflower (Echinacea purpurea) *and butterfly weed* (Asclepias tuberosa), *wood geranium* (Geranium maculatum).

Center: Bergamot (Monarda fistulosa), *aster spp., dahlia.*

Bottom: Purple toadflax (Linaria purpurea), *yellow coneflower* (Ratibida pinnata) *with bergamot.*

Biennials

Biennials are plants that grow for one season, overwinter, bloom, and set seed the next season, then die, completing their biological lifecycle. There are a few excellent bee plants and garden subjects in this category that are very worthwhile. The angelicas are an excellent group of biennial plants for bees and many other beneficial insects. The short and densely clustered floral tubes in the umbel flowers provide easy access to nectar and pollen. Because there are many clustered flowers in each umbel, bees often move over them slowly, making these flowers ideal for bee watching. *Angelica stricta* 'Purpurea' (for zones 5 to 11) is a favorite angelica of many. It has striking upright form to 4 to 5 feet tall, burgundy foliage, and pink flowers. It needs soil moisture and prefers good soil fertility. Though it dies after it blooms, it produces a lot of seed and usually will reseed itself, generating wonderful patches of this productive plant. Small bees and honeybees favor its flowers. Some angelicas are perennial.

A biennial very attractive to carpenter bees is clary sage (*Salvia sclarea*). The entire plant is robust, with large, distinctive, fuzzy leaves and a huge elaborate flower spike with pink or violet calyxes. Carpenter bees are strong enough to pry open the large, bilaterally symmetrical flowers; they leave with evidence of their floral visits in the form of a deposit of distinctive white pollen on their shiny backs. A favorite cultivar with showy violet calyxes is called 'Piemont'.

Another group of good biennials are the mulleins (*Verbascum*), though some are perennial. A favorite is silver mullein (*Verbascum bombyciferum*). It has white, felted foliage and bright yellow flowers on distinctive candelabras. Greek mullein (*Verbascum olympicum*) is a giant that grows to 5 to 8 or more feet with huge candelabra spikes of flowers in bright yellow. The flowers are small, but closely clustered, making an impressive show bees also love. It spends its first year as a rosette of attractive, strappy, gray leaves. It will reseed after bloom if you let the spikes mature and dry. Both plants require well-drained soil and sun.

Foxgloves are biennials many are familiar with. Though not native to North America, foxgloves have naturalized in many areas. Bumblebees are a common pollinator, and anthers on the top of the flower dust the back of each bee with pollen as bees seek nectar held deeply in the corolla.

Below, left: Angelica stricta *'Purpurea' is a striking biennial in leaf and flower.*

Below, right: Greek mullein (Verbascum olympicum) *is a robust and statuesque plant that reaches 7 to 8 feet in height.*

Cactus

There are a number of cacti that grow easily in different regions of the country and make dramatic garden specimens, whether in bloom or not. Our favorites are the prickly pears (*Opuntia*). We usually think of cactus as only living in hot, dry climates, yet there are a number of prickly pears that endure freezing, and others that thrive in humid climates. They have large, bowl-shaped flowers with a profusion of anthers that attract a number of specialist bees. The blooms are extremely showy and abundant and come in brilliant hues of yellow, pink, red, or burgundy, to just name a few. There are many different types of prickly pears, from varieties that grow 10 feet tall to ground-cover forms. Some have purple pads, and some will grow in snowy mountains. Some have a multitude of spines, others are spineless. The flower colors and fruit colors vary immensely. The common trait is they are very easy to grow and propagate. To propagate, just break off or cut a pad at the base and bury it halfway into the soil.

Prickly pear (Opuntia) cactus in bloom. There is a large variety of prickly pear cacti, some with opulent and brilliantly hued flowers.

Some *Diadasia* bees are pollen specialists (oligoleges) of cactus, though many other bees visit the cactus flowers as well. *Diadasia* are ground-nesting bees that often nest in aggregations on bare ground, which is often found in cactus gardens. An excellent book on garden design, *Plant-Driven Design* by Scott Ogden and Lauren Springer Ogden, details a number of ways to grow cactus (along with much other information) in different environments and in a more naturalistic way than the usual rather stark cactus garden. For a list of additional bee-friendly cacti, see the Southwest Region Plant List on page 194.

Prickly pear cactus flower with bee inside.

Shrubs

Shrubs add structure and dimension to your garden both in summer and winter. Foliage or bark can be very handsome even when the plant is not flowering. Shrubs come in many shapes and sizes. Some are deciduous—losing their leaves in winter—while others are evergreen. Some have vertical form, while others are strongly horizontal. Size needs to be carefully noted because some attain the size of small trees, while others remain very small. Any plant in a too-small space will require frequent, time-consuming maintenance and can crowd out other desired plants, especially in a small garden. Because shrubs are often slower growing and generally more expensive than annuals

Chitalpa (× Chitalpa tashkentensis) *is a shrub that thrives in hot climates and has large, fragrant, orchidlike blooms for many months each summer.*

or perennials, it is best to research possible choices to make sure each makes sense for a particular location.

Some deciduous types with handsome form or bark, which is a beautiful feature in any season, include the bee-friendly 'Natchez' crape myrtle (*Lagerstroemia indica* × *fauriei*). The larger manzanitas (*Arctostaphylos*) have leathery, silver leaves and deep red, smooth bark on branches that are dramatically muscular, bringing to mind an athlete in motion. A couple of fast-growing, long-blooming, fairly tender shrubs with very showy flowers are the desert willow (*Chilopsis linearis*) and tree mallow (*Lavatera thuringiaca*). The desert willow is either grown as a shrub or shrubby tree. It has fragrant, orchidlike blooms all summer.

Many shrubs have a specific season of bloom—winter, spring, summer, or fall. Shrubs provide a large area of floral resources for bees during each plant's specific bloom period. For example, manzanitas and wild lilac (*Ceanothus*) often provide critical early season resources for bumblebee queens. Some shrubs will need pruning in winter. Shrubs such as chaste tree should have their spent flowers removed in summer if time permits as they may rebloom if this is done, while others, such as the redbuds, will drop their flowers when bloom is finished. Some shrubs' flowers are also attractive to beneficial insects, while others have berries that attract birds.

Escallonia, a plant commonly used for hedges, has white or pink flowers highly attractive to bees and often butterflies. The buckthorns—both the Carolina buckthorn (*Rhamnus caroliniana*) and, in the West, coffeeberry (*Rhamnus californica*)—are highly sought after by beneficial insects and bees, while birds enjoy the fruits that follow. Yellow twig rabbitbush (*Chrysothamnus nauseosus*) provides bees of many varieties with pollen and nectar late in the season. Its yellow stems light up gardens in the winter. The widely adapted three-leaf sumac (*Rhus trilobata*) has flowers attractive to bees, then showy fruit

*Facing page: Chaste tree (*Vitex agnus-castus*).*

clusters and good fall color. The flowers of cotoneasters are highly favored by honeybees, and the berries are sought after by birds in the fall. Wild lilacs (*Ceanothus*) are an important early pollen source for many native bees, and birds then feed on the seeds later in the season. The bees enjoy both the dainty white flowers and the blueberry-like berries of the serviceberry (*Amelanchier* spp.).

BEE-FRIENDLY SHRUBS

A few good examples of bee-friendly shrubs are:

- Chaste tree (*Vitex agnus-castus*)
- Chitalpa (× *Chitalpa tashkentensis*)
- Coffeeberry (*Rhamnus californica*)
- Cotoneaster (*Cotoneaster*)
- Desert willow (*Chilopsis linearis*)
- Escallonia (*Escallonia*)
- Grevillea (*Grevillea*)
- Hebe (*Hebe*)
- Holly (*Ilex*)
- Hydrangea (*Hydrangea*)
- Mahonia (*Berberis*)
- Manzanita (*Arctostaphylos*)
- Mock orange (*Philadelphus*)
- *Prunus* spp.
- Redbud (*Cercis*)
- Saint-John's-wort (*Hypericum prolificum*)
- Serviceberry (*Amelanchier*)
- Strawberry tree (*Arbutus unedo*)
- Summersweet (*Clethra alnifolia*)
- Sweetspire (*Itea*)
- Toyon (*Heteromeles arbutifolia*)
- Wild lilac (*Ceanothus*)

Hydrangeas have both fertile and sterile flowers, but many popular hydrangeas found in gardens have been bred to favor the large sterile flowers. A better choice is Hydrangea aspera, *whose almost entirely fertile flowers are fragrant and highly attractive to bees.*

Facing page, from left to right,

Top: Coffeeberry (Rhamnus californica), *strawberry tree* (Arbutus unedo).

Center: Mock orange (Philadelphus), *wild lilac* (Ceanothus), *Saint-John's-wort* (Hypericum prolificum).

Bottom: Manzanita (Arctostaphylos), *summersweet* (Clethra alnifolia).

Trees

When people think of bee plants, they typically envision flowers, not trees, but trees are the ultimate plant for generating a very large area of floral resources in your garden. They also can be a significant landscape feature around a home and are very useful in casting shade to ameliorate summer temperatures. Many provide secure nesting and perching sites for birds. Those with berries can provide significant food for them. Some trees may be only 10 feet tall, while others attain heights well over 100 feet. Careful consideration is obviously key to a tree being an asset rather than a liability in one's yard. Some grow extremely slowly, while others are fast growing. In general, fast-growing trees tend to have weaker wood and can break more easily. Bees also use many fruit trees, in part because of their abundant bloom (see page 110).

Facing page, from left to right,

Top: Arbutus *'Marina'.*

Bottom: Black locust (Robinia pseudoacacia)*, flowering crab apple* (Malus)*, linden or basswood* (Tilia)*.*

BEE-FRIENDLY TREES

Some good bee-friendly trees are:

Small

Acacia (*Acacia*)*

Chokecherry (*Prunus virginiana*)

Flowering crab apple (*Malus*)

Hawthorn (*Crataegus*)

Japanese maple (*Acer palmatum*)

Magnolia (*Magnolia* spp.)

Palo verde (*Cercidium*)

Plum (*Prunus* spp.)

Snowbell shrub (*Styrax*)

Sumac (*Rhus* spp.)

*Check size before purchase. Some acacia species and cultivars grow larger.

Medium and large

American linden or basswood (*Tilia americana*)

Arbutus 'Marina'

Black locust (*Robinia pseudoacacia*)

Chinese tallow tree (*Sapium sebiferum*)

Holly (*Ilex*)

Magnolia (*Magnolia* spp.)

Mesquite (*Prosopis*)

Red maple (*Acer rubrum*)

Sourwood (*Oxydendrum*)

Tupelo (*Nyassa sylvatica*)

Willow (*Salix*)

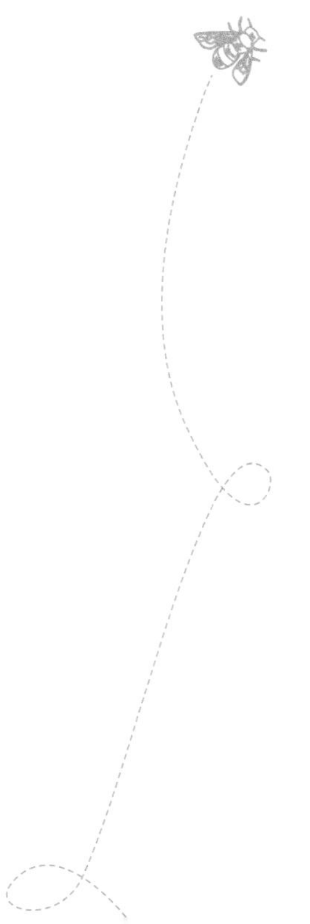

1715

NATIVE PLANTS

Research has shown that native bees usually prefer native plants over nonnatives, although many of our native bees are generalists and forage from a variety of nonnative species as well. Native plants from the regions in which the bees originated are the best choices for bee gardens for a variety of reasons. Native plants also support many native birds and insects better than nonnative species, particularly native moth and butterfly species. Moths and butterflies also act as pollinators, and their caterpillars are an important food source for fledgling birds. Native plants are usually less attractive to insect pests, are adapted to the soil and climate of a region, and often require far less fertilizer or soil preparation than nonnative species do. Importantly, their appearance also expresses a sense of place often lacking in our landscapes. Try to include at least 50 percent native plants in your garden.

There are many excellent native plant nurseries that supply native plants that will thrive in each specific area of the country. Most nurseries also include good plant cultural information along with plant descriptions that help guide your selection of plants appropriate for your specific situation. See the references in the back of the book for more information and for regional plant lists.

EARLY AMERICAN HONEY PLANTS

Frank Chapman Pellett was born in 1879 and wrote thirteen books on bees, wildlife, and horticulture. His book, *American Honey Plants*, about bee forage, was first published in 1920 and went through four editions. Despite the book's age, its information is still vitally important to us. He details the importance and potential of wild plants in each state in providing honeybees with forage and delineates both major and minor forage plants in different seasons, weather, and climates. He was concerned as early as 1907 that many of the wildflowers that were abundant during his childhood were disappearing.

"Changing conditions are rapidly removing one plant and substituting another in many sections. The clearing of land and planting it with cultivated crops is rapidly curtailing bee range, as no cultivated crops that are being planted are equal to the desert flora which is being removed." He also describes the importance of a varied diet of pollen and nectar for honeybees: "The ideal situation for beekeeping is where there are at least three plants which yield surplus nectar in considerable quantity, and which bloom in different periods. Besides these main sources, there should be a great variety of minor plants yielding both pollen and nectar throughout the season to support the bees between the main flows."

In his book, Pellet describes the agricultural plants, garden plants, and native plants in each state—annuals, perennials, shrubs, and trees—that comprise bee forage. A browse through his book is to touch the landscape of the past. From analysis of alfalfa varieties from Turkistan, Peru, Arabia, and the Baltics, to our native wildflowers—*Amsinckia* or fiddleneck, goldenrods, asters, milkweeds, gaura, lupine, wild onion, prairie clover, giant hyssops, and more—he extols the virtues of native plants across the nation and their capacity to support bees with pollen and nectar. Many of these plants are still good subjects in our gardens.

SUPERBLOOMERS

Some plants bloom for more than one season or have a couple of waves of bloom. These are rewarding garden inhabitants both for ourselves and for the bees. Using many superbloomers together will provide a nonstop, easy-to-care-for show with more blooms than foliage and will be filled with bees of many varieties, hummingbirds, and butterflies.

Superblooming plants are usually annuals or perennials, though some shrubs like *Grevillea* 'Ruby Clusters' bloom for six months or more, and mock orange (*Philadelphus*) shrubs bloom profusely and smell sweetly for well over a month in early summer. The strawberry tree also has a prolonged bloom season in fall and is much visited by bumblebee queens and hummingbirds. In contrast, black locust trees and plum trees bloom abundantly for only two to three weeks. While the sheer volume of bloom on trees offer bees huge quantities of forage, when their bloom is finished, they need to be followed by other blooming plants so that bees have a continuous supply of pollen and nectar. Superbloomers, on the other hand, make it easier to keep your garden in continuous bloom.

In dry climates in zones 8 to 10, some easy-to-grow and long-blooming perennials that look great together are 'Purple Haze' hyssop, calamint (*Clinopodium nepeta*), blanket flower, catmint, gumweed, *Aster* × *frikarti* 'Monch', *Salvia* 'Mystic Spires', California buckwheat (*Eriogonum fascicularis*) or sulphur-flower buckwheat, autumn sage (*Salvia greggi*), mealycup sage (*Salvia farinacea*), indigo woodland sage (*Salvia verticillata*), and Iranian wood sage (*Teucrium hircanicum*)—among others.

Many *Agastache*s are hummingbird plants, but the anise hyssop (*Agastache foeniculum*) hybrids or licorice mint hybrids are bee favorites and excellent garden subjects; they have abundant and showy blue-purple spikelike inflorescences for months. The upright flower spikes of the hybrids require little deadheading and the blue-purple colors combine well with many other flowers. Some good hybrids are 'Purple Haze', 'Black Adder', 'Blue Blazes', 'Blue Fortune', and 'Astello Indigo', but more good hybrids and selections are available each year.

Blanket flower (*Gaillardia* 'Oranges and Lemons') is a cheerful perennial in shades of orange and yellow that is another superbloomer. It looks great together with *Agastache* plants, as it does with the *Salvia* 'Mystic Spires' and the mealycup sage, an upright, very easy to grow salvia with deep blue flowers. There are a number of excellent blanket flowers for zones 4 to 10 that grow well in heat, sun, and poor soil, but the cultivar 'Oranges and Lemons' is a bee favorite. Avoid the doubled flower varieties. These plants can be grown as annuals in colder climate zones.

Macedonian scabious (*Knautia macedonica*) has deep burgundy flowers on a plant that is about 1 foot high, but spreads to 2 feet. It prefers well-watered conditions and well-composted soil; when given this situation, it will bloom all summer. It combines well with the lower-growing goldenrods, such as 'Wichita Mountains' or the pale yellow 'Little Lemon'. As a pastel contrast, *Aster* × *frikarti* 'Monch' has light violet-blue daisy flowers with yellow centers and is a tidy and well-behaved garden subject. Removing spent flowers prolongs its bloom, as it does for many other plants.

Catmint and calamint both have a billowy form, and the plant forms and colors are soft. The foliage and flowers of catmints lie on the ground, while the calamints' white blooms froth upward. Some people trim the first flush of bloom from the catmints, but this is not necessary. The calamint needs no care from first bloom in late June until freezing weather finally defeats the flowers.

DEADHEADING

Removing spent flowers often stimulates plants to rebloom, or keep blooming. The big question for most people is how much to cut the plant back. There are several ways to approach this question. In some plants like catmints, after the spring bloom, the foliage and flowers decline and can look unattractive. If you look closely, new, fresh foliage is evident at the center of the plant. This indicates the plant will benefit from having this old, declining foliage removed, allowing the new foliage to grow unimpeded. There are many plants of this type—all indicating that they are ready for renewal by growing fresh, strong young growth at the center of the plant.

In other plants, like the evergreen penstemons such as the Rocky Mountain penstemon or perennial mulleins, flowers are upright spikes rising from an attractive basal rosette of leaves. In this case, the entire spike should be removed when it is finished. Other, more bushy penstemons can be sheered after bloom is finished. This usually stimulates the plant to produce more spikes. The upright daisy flowers of purple coneflowers can be removed as well.

Other plants, like some of the large sages, just have spent flowers removed down to a strong bud, as one would do with a rose in summer to stimulate more bloom. The general rule is, if the plant's spent flowers and foliage are obviously declining, look ugly, and are sparse, or the plant is just a mass of spent flowers, cut them. If the foliage looks good, just cut off the spent flowering spikes. It is helpful to consult a book on perennials to help guide you in this process.

The downside of deadheading is that some plants, like black-eyed susans, evening primroses, purple coneflowers, asters, coreopsis, and perennial sunflowers, have flowers with seeds birds like goldfinches feed on. Removing them removes the bird's food. Plants with hollow stems, like teasels, are places where cavity-nesting native bees will nest. Cutting them down after they die in mid- to late summer removes potential nesting sites or, in the worst case, the overwintering bees that would emerge next year. There is obviously a happy compromise: you can leave all the mid- to late season spent blooms on plants you know birds will feed on. This way you also have fall and winter interest from the often interesting seed heads.

Facing page: Many perennials, including the mullein Verbascum nigrum *'Album', benefit from having spent flowers removed to promote prolonged flowering periods.*

Valley gumweed (*Grindelia camporum*) is a western wildflower with bright yellow daisy flowers that will thrive in other warm and dry areas, though it shows its appreciation for regular summer water by blooming all summer without pause. Both native bees and honeybees visit its blooms. Another yellow daisy that blooms nonstop all summer is the brown-eyed susan (*Rudbeckia trilobata*). It is a short-lived perennial from the Iowa prairie that can be grown as an annual and absolutely should be used more widely. It grows in wide variety of climates. Some grow it as a biennial.

Though a nonnative plant and not a sage, Russian sage (*Perovskia atriplicifolia*) blooms all summer with delicate and profuse spires of blue that combine well with yellow flowers. It grows well at low elevations and is practically shrublike in the infertile, granitic soils of the Sierra Nevada mountains and the Rocky Mountains.

Another group of plants that can tolerate drought and, with some species, mountain conditions, are the wild buckwheats. A species that blooms for practically twelve months is white-flowering California buckwheat (*Eriogonum fasciculatum*). It grows best in the dry West and is covered in bloom on a tidy plant. Another long-flowering wild buckwheat that grows from the West to the Rocky Mountain region is the sulphur-flower buckwheat (*Eriogonum umbellatum*). It is low-growing, long-lived, and forms a dense mound covered with yellow flowers only a foot high. Bees, butterflies, and beneficial insects visit the wild buckwheats, and there are many more good species.

Some examples of summer annuals that bloom all summer are cosmos, both the pink or white cosmos and the orange Klondike cosmos. Both benefit from deadheading. Spider flower (*Cleome*), about the same size as the cosmos, also blooms all summer. It is from South America and comes in white, pink, or purple and can be combined with pink or white cosmos for a graceful picture. There are also native plants of the genus *Cleome*, such as the Rocky Mountain bee plant.

Multiheaded sunflowers bloom for a much longer period than the single-headed varieties, and some, like the Japanese sunflower and wild sunflowers, will bloom from midsummer through frost. Some smaller plants that are easy to grow and bloom all summer are basil,

Facing page: Superblooming annual multiflowered sunflowers, Russian sage, and sedums growing among the summer blooming chaste trees prolong the bloom period of this flower border.

annual blanket flowers, and bidens (*Bidens ferulifolia*). Bees love basil. There are many varieties of basil available—lemon, lime, cinnamon, Greek mini, and various Thai varieties. If you are lucky, you may be able to find African blue basil, a perennial in frost-free climates. The flowers of Thai varieties are deep purple and very showy. They often don't need cutting back.

SUPERBLOOMING PERENNIALS

Some of the best superblooming perennials for bees are:

Anise hyssop (*Agastache foeniculum*)

Aster × *frikarti* 'Monch'

Bidens (*Bidens ferulifolia*)

Blanket flower (*Gaillardia*)

Calamint (*Clinopodium nepeta*)

Catmint *(Nepeta)*

Cuphea (*Cuphea*)

Dalea (*Dalea*)

Germander (*Teucrium*)

Knotweed (*Persicaria amplexicaulis*)

Lavender (*Lavandula*, some species)

Macedonian scabious (*Knautia macedonica*)

Oregano (*Origanum*, some species) and ornamental oreganos

Penstemon (*Penstemon*)

Pincushion flower (*Scabiosa*)

Purple coneflower (*Echinacea purpurea*)

'Purple Haze' hyssop (*Agastache* 'Purple Haze')

Russian sage (*Perovskia atriplicifolia*)

Sage (*Salvia*, including *Salvia* 'Mystic Spires', *S. farinacea*, *S. darcyi*)

Sea holly (*Eryngium*)

Sunflower (*Helianthus annus*)

Tall verbena (*Verbena bonariensis*)

Valley gumweed (*Grindelia camporum*)

Facing page, from left to right,

Top: Cuphea *x 'Kirsten's Delight', valley gumweed* (Grindelia camporum).

Center: Oregano spp., sea holly (Eryngium), *anise hyssop* (Agastache foeniculum).

Bottom: Blanket flower (Gaillardia *'Oranges and Lemons'*), Bidens ferulifolia *(yellow) and catmint* (Nepeta).

Plants that Attract and Support Bees and Beneficial Insects

A number of plants support both bees and beneficial or predatory insects. Beneficial insects add to the biodiversity of your garden. They prey on pests that feed on plants, helping to keep your garden healthy without the use of harmful pesticides and chemicals. Many, but not all plants that attract and support beneficial insects are plants in the Apiaceae (umbel) family and Asteraceae (daisy) family. The plants in these families have flowers with short, easily accessible nectaries that beneficial insects without specialized mouthparts for gathering nectar are able to access. Ladybugs, parasitoid wasps, soldier bugs, and other predatory insects all benefit from the additional nutritional resources of nectar, which can help them to live longer and lay more eggs.

Angelica stricta 'Purpurea' and most angelicas attract many beneficial insects.

Lacewing and hoverfly, or flower fly, larvae are predatory, while the adults only feed on nectar. They need flowers with accessible nectar to complete their life cycle and lay eggs. Hover flies visit many of the same flowers that bees do in search of nectar. Because their mouthparts are small and short, they gravitate to flowers with short, easily accessible nectar tubes. The larvae, or grub, feed on soft-bodied, small insects like aphids. Carrots, parsley, celery—all biennials—and the annuals dill and cilantro are common examples of the type of plant in vegetable gardens that support these beneficial insects. Angelica, yarrow, and fennel are a few of commonly grown biennials and perennials planted for predatory insects. Bees visit all of them.

Lacy phacelia (Phacelia tanacetifolia) *is a robust annual that prefers cool growing conditions and is widely grown for bee fodder but attracts many other beneficial insects.*

Many native annuals, shrubs, and trees support both pollinators and beneficial insects and are the best plants to use for this purpose. Researchers at the University of California, Davis, and University of California, Berkeley, found that hedgerows of native plants that support beneficial insects around farms also attract and support bees. Some of the main plants used in the studies were manzanitas, *Prunus* spp., wild lilacs, flannel bush (*Fremontodendron*), California buckwheat, elderberry, redbuds, willow, coyote brush, mock orange, coffeeberry, and toyon. Perennials used in the study were native asters, phacelia, California fuchsia, gumplant, goldenrod, yarrow, and milkweeds. Many of these plants make good garden subjects.

In different regions of the country, in differing soils and climates, specific plants are grown for attracting and supporting beneficial insects and bees. Research from the University of Minnesota showed the high potential of a number of native plants to support beneficial insects and bees. The most attractive plants were butterfly weed (*Asclepias tuberosa*), purple coneflower (*Echinacea purpurea*), joe-pye weed (*Eutrochium maculatum*), western sunflower (*Helianthus occidentalis*), false sunflower (*Heliopsis helianthoides*), prairie blazing star (*Liatris pycnostachya*), mountain mint (*Pycnanthemum muticum*), black-eyed susan (*Rudbeckia hirta*), and the goldenrod *Solidago canadensis*. In addition, these plants need far less nutrients than most nonnative perennials, and are far more drought tolerant and resistant to insect pests. Used together, they create a life-filled and floriferous garden display over many months. Some of the milkweeds, goldenrods, and perennial sunflowers can spread over time and may need controlling or careful placement. Their beauty, toughness, profuse bloom and great attractiveness to many beneficial insects and pollinators make them worthwhile garden residents, especially in large gardens.

The Xerces Society (www.xerces.org/pollinator-conservation/plant-lists) and Pollinator Partnership (www.pollinator.org/guides.htm) have good regional lists of plants appropriate to each region. University Extension offices often do as well. And in the Regional Plant Lists in the back of this book (page 190), native plants for each region are marked.

Plants that Attract and Support Bees and Hummingbirds

Hummingbirds flitting around a garden are a favorite sight for many. If you would like to attract both hummingbirds and bees to your garden, you have a number of plant options. Hummingbirds are strongly attracted to plants with red flowers. Red flowers with constricted floral tubes are often cited as hummingbird plants, yet there are many other plants that hummingbirds visit that are not red. Constricted nectar tubes act to exclude many other pollinators from nectar held deep in flowers that is accessible to hummingbirds due to their long tongues. While some large bees have mouthparts that let them reach nectar at the bottom of a medium-length tube, their vision system is different from birds. Some colors like red and pink appear achromatic to bees, meaning they appear little different than gray. However, the copious amounts of nectar produced by these plants appeals to a number of bees. Carpenter bees often rob nectar by making a slit in the long nectar tubes, and sometimes honeybees or other native bees reuse these slits. Some of the flowers are accessed by large, powerful bees like bumblebees or carpenter bees, which have long tongues and can force petals apart in flowers such as lupines, or can access nectar held in long floral tubes like in bergamot (*Monarda fistulosa*). Others, like the smaller penstemons, manzanita, and California fuchsia (*Epilobium)* have nectaries that are also accessible to some smaller bees because they can fit in the narrow or constricted floral tubes.

Carpenter bee making a slit in a tubular flower.

Honeybee using the slit made by a carpenter bee to access the nectar.

Facing page: Salvia forsskaolii, *in the foreground, is an attractive hummingbird plant. It combines in a harmonious symphony of hues with* Persicaria amplexicaulis *'Taurus' and* Salvia sylvestris (nemerosa).

PLANTS THAT SUPPORT BOTH BEES AND HUMMINGBIRDS

Some good examples of hummingbird plants that are also visited by bees include:

Bergamot (*Monarda fistulosa*)

Bottlebrush (*Callistemon*)

California fuchsia (*Epilobium*)

Chitalpa (× *Chitalpa tashkentensis*)

Columbine (*Aquilegia*)

Currant (*Ribes*)

Desert willow (*Chilopsis linearis*)

Fairy duster (*Calliandra*)

Grevillea (*Grevillea*)

Lobelia (*Lobelia*), perennial

Lupine (*Lupinus*), many

Manzanita (*Arctostaphylos*), some

Ocotillo (*Fouquieria splendens*)

Penstemon (*Penstemon*)

Red bird of paradise (*Caesalpina pulcherrima*)

Sage (*Salvia*)

Snowberry (*Symphoricarpus*)

Wild hyssop (Agastache)

Facing page, from left to right,

Top: Penstemon *'Garnet', California fuchsia (*Epilobium *'Catalina').*

*Bottom: Texas sage (*Salvia coccinea*).*

Factors that Influence Bloom, Nectar, and Pollen

There are many plants that provide pollen and nectar for bees, some native, others from different regions around the world. Some plant species are widely adapted and will grow in large geographic areas, while others have specific climate, water, and soil requirements.

Plants that are adapted to the heat, moisture, humidity, aridity, cold, soil type, and pest and disease pressures in your area will thrive and bloom for an optimum period of time. Plants used in conditions they are not adapted to will usually have curtailed periods of flowering. For example, prairie flowers from the Midwest are adapted to the deep, fertile soils and the humidity and moisture found there. If you plant them in infertile, dry, shallow soils with low humidity, the stress will shorten bloom time. Wild indigos (*Baptisia*), joe-pye weed (*Eutrochium maculatum*), black-eyed and brown-eyed susans, and ironweed (*Vernonia*) are a few examples of flowers that bloom for just a couple weeks in hot and dry summer climates, versus for one to two months in the Midwest.

The same plant species from different geographic and climatic areas will have different tolerances and adaptations in terms of soil preferences and tolerance for drought, heat, humidity, and cold. For instance, a California native shrub that bees eagerly visit called coffeeberry grows from coastal regions into the mountains, but nurseries mostly carry selections from cool, coastal regions. These coastal-origin plants often fail to thrive in hot, inland areas. It is possible to find populations of plants from these warmer, drier areas at specific nurseries, but most people don't know to ask. An example of a species that has distinct needs, yet is planted across the United States, is the Eastern redbud, a deciduous shrub that has very showy, magenta blooms in spring that large bees like bumblebees, carpenter bees, *Habropoda* bees, and some *Osmia* bees frequent. Its beautiful leaves and form make it attractive in every season. This plant grows naturally from Texas to Washington, DC, and up into Ontario, Canada. It is hardy to zone 4. When planted in California, however, it forms a scraggly bush prone to sunburn unless grown in excellent soil in cooler areas. The same shrub grown in Washington, DC, is robust and highly showy, with distinctive leaves, form, and presence. The redbud species native to California (*Cercis occidentalis*) thrives in heat, drought, and poor soil and is a glorious sight along some roads in spring, and should be grown in more gardens there.

Infertile soil, drought, or insufficient water limits plants' growth and development. It also limits the number and size of the flowers, as well as length of bloom, and nectar and pollen production. Heat can cause a plant's nectar to become more concentrated and viscose; if the heat is extreme, it can cause the plant to cease production altogether. To determine which bee plants are best for your area, consult the regional lists on pages 190–206, The Xerces Society's regional plant lists, USDA Natural Resources Conservation Service, Pollinator Partnership, Master Gardener or Naturalist programs, or other regional pollinator organizations. Talk to your local nursery, consult books in the Resources section on page 184, and, just as importantly, note what plants are growing well in other gardens near yours that have bees visiting them.

Why Don't I See Bees on My Bee Plant?

You may take a daily afternoon walk past a patch of a dainty, white or yellow, summer-blooming wildflowers called tarweed (*Hemizonia* spp.), never see any bees on it, and assume that it is not a good bee plant. But take a walk at eight in the morning, and you may be surprised to see the small flowers covered with native bees, many carrying loads of white pollen, so copious in number that you can hear them.

You may become familiar with the plants bees visit in your garden, only to be confounded when you go to a friend's or neighbor's house and see either no bees on the same plants, or bees on plants that they don't visit in your own garden. A friend has a lot of deer grass, a large, very striking native grass native to the western United States, in her garden and never saw any bees on it. Her colleague at work described how a large number of bees were collecting pollen from the same plant at her house. In the same way, we never see any bees on the native reeds species that grow around a small pond nearby, yet a friend reported that there were so many bees collecting pollen from her native reeds and sedge species around her pond, she thought someone must have put a hive near it.

This is likely related to the size of a foraging patch (see page 120) and the other plants available in the area. Native bees and honeybees have preferred plants to forage on for pollen. In the same way that we have our preferred restaurants, bee-foraging behavior depends on the best, preferred offerings *and* quantity of those and other plants in the neighborhood. Based on these parameters, bees will visit particular plants and not others. As the quality and quantity of forage changes from place to place and over time, so does the pattern of foraging behaviors. When there is little forage available, bees will visit plants they might not otherwise use.

Always observe the activities of bees in your garden and neighborhood at various times of day before you draw any conclusions about whether a plant is attractive to bees, because plants bloom at different times of the day or night, catering to their insect or bird pollinator's period of activity. The timing of release, duration, and composition of nectar and pollen is determined both by genetics and by the environment. Plants of the same species usually produce nectar and pollen at similar times and for similar lengths of time (duration), although there can be individual variation. Factors such as heat and aridity also affect nectar

Facing page: Honeybee on bachelor button.

or pollen production. During the course of the day, you may observe bees on a plant's flowers in the morning and not in the afternoon because some plants release pollen or nectar only in the morning before temperatures rise. Thus, as you walk through your garden at different times of the day, and in different weather, you may see differing numbers or no bees at all visiting flowers.

A few examples of flowers that release floral rewards mainly in the morning are many poppies, mulleins, and spider flowers. Bees are clearly very active in the mornings collecting floral rewards from these flowers. In poppy flowers, bumblebees move swiftly in a characteristic round and round movement through the feathery anthers collecting pollen, a behavior called wallowing. In the morning, the activity appears frantic. In the afternoon, the flowers may have little activity, reflecting a lack of pollen production.

A native bee that is active very early in the morning before the honeybees are out is the squash bee (*Peponapsis*), a specialist bee that only visits flowers in the squash family. Male squash bees sleep in the flowers and become trapped when the blooms close at night. If you visit your garden early in the morning, you will likely see them in the flowers. You can actually gently squeeze the closed flowers and if they buzz, you'll know a squash bee is in there. Female squash bees are ground-nesting bees and prefer to nest close to the squash plants. Make sure to leave some untilled soil around your vegetable garden area for this purpose.

Plants That Are Bad for Bees

Not every plant has floral resources bees can recognize, utilize, or access. For instance, some sunflowers have been bred to be pollenless, depriving bees of an important source of protein. Some plants may not produce pollen or nectar, while others have so many petals that the anthers and nectaries are not accessible. Double roses are an

Facing page: The flowers of Cleome 'Violet Queen' open early in the morning and close in the heat of the day.

often-cited example, but some dahlias, camellias, blanket flowers, black-eyed susans, and sunflowers can have so many petals no access is possible. These flowers have been selected for their profuse petals and form to appeal to humans, without regard to pollinators.

Still other flowers may be pollinated by flies, moths, bats, or birds. Flowering tobaccos are pollinated by night-flying moths. Evening primroses have nectar only accessible to moths, but bees avidly gather the often sticky, long strands of pollen. As mentioned in the discussion about hummingbirds (page 88), plants with the red-to-orange flowers with long nectar tubes cater to hummingbirds and sometimes exclude other pollinators. Yet, carpenter bees and bumblebees actively slit the nectar tubes with their powerful mandibles and steal the nectar without pollinating the plant. Other bees then often rob nectar through these slits.

Many of the plants that you see recommended for use near swimming pools are not candidates for bee gardens. Because pool owners don't want live or dead bees in their pool, most of these plants have been selected specifically because bees don't visit them. These plants may not flower, as in the case of ferns, or their blooms may not have accessible or recognizable nectar for bees.

Facing page: The single dahlias and roses on the left are excellent bee plants, while the double dahlias and roses on the right are not bee-friendly, because their abundance of petals prevents bees from accessing the pollen and nectar.

PLANTS NOT ATTRACTIVE TO BEES

Some common plants that are not attractive to bees include:

- Bigleaf hydrangea (*Hydrangea macrophylla*)
- Canna (*Canna × generalis*)
- David viburnum (*Viburnum davidii*)
- Daylily (*Hemerocallis*)
- Ferns
- Fig, edible (*Ficus*)
- Fortnight lily (*Dietes*)
- Gazania (*Gazania*)
- Grasses, nonnative
- Hop bush (*Dodonea*)
- Japanese laurel (*Aucuba*)
- Lily of the Nile (*Agapanthus*)
- Maiden grass (*Miscanthus*)
- Peruvian lily (*Alstroemeria*)
- Red hot poker (*Kniphofia*)
- Sago palm (*Cycas revoluta*)

Although the pollen or nectar of most plants does not pose a problem for bees, a few plants have nectar or pollen that is mildly or strongly toxic to bees or their larvae. Toxicity susceptiblity varies in each plant species in its effects on native bees and honeybees. Toxicity effects also vary across native bee species. Much is unknown about plant pollen and nectar toxicity to native bee species.

Several native plants are toxic to honeybees or their larvae, including Death camas (*Zigadenus* or *Toxicoscordion venenosum*), California cornlily (*Veratum californicum*) and California buckeye (*Aesculus californica*). In the case of the California buckeye, the nectar causes deformed larvae in honeybees. The bloom period is only about two to three weeks long, so damage from nectar will not overwhelm the hive, especially if there are other floral resources for bees to visit. These plants are rarely planted in urban gardens. In the southeastern United States, a very attractive native shrub, swamp titi (*Cyrilla racemiflora*), causes purple brood, a condition that kills honeybee larvae. As it has a limited flowering period, commercial beekeepers move their hives from the vicinity during bloom when many plants are present. The Carolina jessamine (*Gelsemium sempervirens*) is visited by both honeybees and native bees. Native bees are not affected by its nectar, but according to some reports honeybees may be. Though the flowers are sweet smelling, the whole plant is toxic, and if children are present you may want to avoid the plant for this reason.

Some lindens or basswoods (*Tilia*), such as the European littleleaf linden (*T. cordata*), are reportedly toxic to some native bees in some conditions, such as aridity. The toxicity can vary across bee species.

Some plants produce nectar that makes toxic honey. Mountain laurel (*Kalmia latifolia*) is one of these.

Facing page: It is easy to see that the shrub Hydrangea aspera *is a bee favorite by the multitudes of honeybees that visit its blooms.*

Keeping Track of Bee Plants

It is a rewarding experience to keep a notebook throughout the year and write down the names of plants and which bees are visiting them. This is a fun and informative exercise wherever you go—from your home, to visiting friends, to walks around the neighborhood, or anywhere you go in the world. Much information can be gleaned this way, much discovered and much shared. Everyone can be a local expert.

CHAPTER 3

bee-friendly plants for edible gardens

Vegetable gardens are an ideal place to include bee-friendly plants. Many of our food plants require bee pollination, including squash, melons, cucumbers, fruit, and some nuts such as almonds. Bee-pollinated fruits are often more numerous and larger.

A number of plants are self-fertile—tomatoes, eggplants, peppers, and tomatillos—which means the plant can accept its own pollen and doesn't need another plant. However, pollen from other plants often produces larger, higher-quality fruits than when the plant self-pollinates. Also, these self-fertile flowers need to be vibrated to move the pollen from the male anthers to the female part of the flower, the stigma. Although this can be done by the wind, bumblebees are far more efficient. With their powerful bodies and wings, bumblebees are able to vibrate pollen from high up in the flower, a behavior honeybees lack. Bumblebees are often used in commercial greenhouse operations for this purpose. Squash bees are specific visitors to cucurbits (the gourd family, including melons and cucumbers, as well as squash), but many other bees, including honeybees readily visit them as well.

Native bees are very efficient pollinators, in part, because some are active early in the morning and later in the evening when it is too cool or rainy for honeybees to leave the hive. Research has shown that small farms next to natural areas often can receive all their pollination requirements from native bees. Honeybees gather just pollen or just nectar on each foraging trip, so they may or may not come into contact with both the male and female reproductive parts of a plant, depending on the configuration of the flower.

*Previous spread: This vegetable garden has many bee-friendly plants incorporated in it. Pictured here are bright orange Iceland poppies (*Papaver nudicaule*).*

Facing page: Single dahlias and yellow evening primroses are cheerful and bee-friendly companions to trellised cucumbers.

When flowers do not receive adequate bee visitation, fruits such as watermelon, squash, strawberries, tomatoes, and peppers may be small or misshapen. Proper development of fruits is based on seed development. Seed development is a function of pollen grains successfully being transferred to plant stigmas. Although other insects may visit vegetable flowers, bees' hairy bodies readily attract and carry pollen. According to the University of Minnesota Extension, each flower may need ten to twelve visits by bees, and research has shown that the number of separate bee visits is what is important, not how long each bee spends in the flower.

A tidy hedge of bee-friendly Greek basil in front of a line of peppers in a vegetable garden.

Pollination is not necessary for some garden vegetables, such as leafy greens and root crops, but they require bee pollination in order to produce seed stock for planting. Honeybees are used extensively for this purpose in commercial seed production. Susan Ashworth's excellent book on seed saving, *Seed to Seed: Seed Saving and Growing Techniques for Vegetable Gardeners*, is a must if you are interested in saving seed from your vegetable plants.

Towering sunflowers and purple spider flowers form a colorful and bee-friendly background to exuberant pumpkins.

Facing page: The giant blue thistle flowers of the cardoon (an Italian vegetable closely related to artichokes) are dramatic in size and form and a honeybee favorite.

Annual Flowers in the Vegetable Garden

Flowering plants are invaluable in vegetable gardens for pollination, natural pest control, and the beauty they create. Vegetables require loose, fertile soil. These same conditions favor many annual flowers, making them ideal companions in vegetable gardens. A number of them are bee-friendly and cheerful additions to an otherwise green scene. Perennials can also be incorporated in vegetable gardens, but you may prefer to keep vegetable gardens entirely annual because slugs, snails, pill bugs, and earwigs—insects that often feed on young vegetable seedlings—can accumulate in the shelter that perennial plants provide.

There are several ways to include annual flowers in your vegetable garden. One method is to incorporate flowers in the beds with the

vegetables. Another method is to include flowers at the end of every row, creating an easy to manage, colorful scene. The ends of the rows are convenient if you are cycling vegetables in and out of the beds frequently and are an easy way to observe bees.

Herbs

Many culinary herbs are bee-friendly. There are annual and perennial herbs, and they come in great variety of foliage, form, flowers, and uses. Some people make gardens devoted to them, while others mix them into the rest of the landscape. Many are very attractive, and they can create lovely visual and fragrant surroundings.

Annual Herbs

Annual herbs have much the same needs as annual flowers or vegetables; they require fairly good soil and water for optimum growth. Some are frost hardy, while others are only grown in summer.

- Anise hyssop (*Agastache foeniculum*)
- Basil (*Ocimum basilicum*)
- Borage (*Borago officinalis*)
- Cilantro (*Coriandrum sativum*)
- Marjoram (*Origanum marjorana*)
- Za'atar (*Origanum syriacum*)

Perennial herbs

There are many perennial herbs that make both useful and attractive garden subjects. Entire books are devoted to them and, for an overview of the vast array of possibilities, a book or a good online herb nursery such as *Mountain Valley Growers* (www.mountainvalleygrowers.com) should be consulted. A garden can be composed of perennial herbs alone, or they can be included in other plantings. Herbs of Mediterranean origin, such as rosemary, lavender, sage, thyme, oregano, marjoram, and fennel require well-drained soil and

Of all the berries, raspberry flowers are honeybee favorites. Crevice dwelling native bees use the old stems for nesting.

some are intolerant of high humidity. Hardiness varies among varieties. They are all very bee-friendly and especially popular with honeybees, which can often be found in the dozens on each plants' blooms. Other commonly grown herbs are of European origin. Together, these herbs bloom over a long season.

Angelica (*Angelica*)
Chives (*Allium schoenoprascum*)
Comfrey (*Symphytum spp.*)
Fennel (*Foeniculum vulgare*)
Hyssop (*Hyssopus officinalis*)
Lavender (*Lavandula*)
Lovage (*Levisticum officinalis*)
Mint (*Mentha spp.*)
Oregano (*Origanum*)
Rosemary (*Rosmarinus officinalis*)
Sage (*Salvia officinalis*)
Thyme (*Thymus*)

Berries

Blackberries, boysenberries, cranberries, loganberries, raspberries, and Saskatoon berries all require insect pollination for good set of large, well-formed berries. Red raspberries and blackberries are self-pollinating, but research has shown that fruits are larger with bee visitation. Bees of many varieties relish the nectar of these plants, especially raspberries, and the hollow stems of old raspberry canes can be nesting sites for some cavity-nesting bees. Honeybee hives are often placed near big patches of blackberries when they are in bloom. Bumblebees are the preferred pollinators of blueberries as they vibrate flowers resulting in larger, more abundant berries. Honeybees are less efficient pollinators because they lack the ability to buzz flowers. In the Southeast, a native bee resembling a bumblebee, called the southeastern blueberry bee, buzz-pollinates blueberries by vibration, but other native bees readily pollinate blueberries as well. Strawberries can be wind pollinated, but studies have shown that bee-pollinated strawberries are redder, have fewer deformities, are more firm, and have a longer shelf life.

Fruit Trees

Many fruit tree flowers require bee visitation for pollination, including apricot, cherry, apple, some pears, nectarine, peach, and plum. Almond trees also require bee visitation for pollination. Some fruit trees such as sour cherries are self-pollinating or self-fruitful, but many fruit varieties are not self-compatible and require cross-pollination from a compatible cultivar that flowers at the same time. For example, Dave Wilson Nursery (www.davewilson.com), a commercial fruit tree nursery that supplies trees to many retail outlets, recommends that one of their top apples, a variety called Akane, be planted with any of the following varieties for pollination: Fuji, Gala, Granny Smith, and Golden Delicious. Some fruit trees are self-fruitful, but will have a more abundant crop with bee pollination. Many tree nurseries and University Extension information bulletins recommend what pollinator varieties should be planted with specific cultivars. Other fruit trees such as Bartlett, Hardy, and Comice pears do not require pollination as they are parthenogenic (fruit will set but is seedless unless pollinated by a bee).

If you have bee-friendly plants in and around your garden, chances are that there will be bees present to pollinate your fruit trees. Both honeybees and native bees readily visit fruit tree blooms, though native bees are the most efficient pollinators. On each foraging trip, native bees may collect both pollen and nectar, and more readily move among varieties, resulting in more contact with male reproductive parts of flowers. They are also out earlier, later, and in more inclement

Orchard mason bees are very efficient pollinators of many fruit tree varieties, such as this apple tree.

weather. Honeybees are used commercially as the main pollinators of large orchards because it is easy to move in huge numbers of colonies needed to pollinate fruit trees during the short period of fruit tree bloom. Each colony contains upward of twenty thousand individual bees, so what they do not provide in quality, they make up for in quantity. Orchard mason bees (*Osmia lignaria*) can be used for pollination in small or home orchards as they are super-efficient pollinators and may be easily attracted to nest in bee blocks. They emerge early in the spring at a time when many fruit trees are blooming and weather may be unsettled. It only takes about 250 orchard mason bees to pollinate one acre of apples because they are active early and late in the day and move between pollinator trees and fruiting varieties better than honeybees. It takes at least one colony of honeybees to accomplish the same pollination.

There are a number of ways bee-friendly plants can fit into an orchard setting or under or around fruit trees. Because each fruit variety blooms for just two to three weeks, you will need other bee fodder to sustain bees for the rest of the season. This is less of an issue for individual species of native bees, which generally are only flying for several weeks. However, these native bees can be helped by supplemental bloom right before and after the fruit tree bloom—especially in years with unusual weather that might lead to a mismatch between flower opening and bee emergence. Some people have bare ground under trees for ease of maintenance. Fruit trees typically require regular maintenance such as winter pruning, summer pruning, thinning, harvest, and spraying during the dormant season, therefore good access is necessary. There are some good low-growing bee-friendly plants that grow well under fruit trees and can tolerate some foot traffic and activities like raking. Creeping thymes are a good choice and come in a variety of flavors like lemon, caraway, lime, orange, as well as cultivars with a number of flower colors from white to red. You can grow several to create a delicate and scented carpet. Creeping marjorams also form good ground covers under fruit trees because they are low, dense, can tolerate some shade, and can take some raking. The variety *Origanum* 'Betty Rollins' is a very attractive cultivar. Bugleweed (*Ajuga*) and heal-all (*Prunella*) are two ground covers that can tolerate some walking on and will also grow in partial shade. A couple of agricultural plants that can be grown on a large or small scale are bird's foot trefoil (*Lotus corniculatus*), a tough, very low-growing legume with yellow pea flowers, and the perennial lawn clovers: New Zealand white or strawberry clover. Both can be seeded very inexpensively and grow in many regions of the country. The clovers benefit the trees by fixing nitrogen and over time will improve soil fertility. Clovers have an abundance of nectar for bees.

*Bee-friendly catmint (*Nepeta spp.*) planted under olives and yucca helps form an all blue/gray scene.*

Wildflowers are also good companions in a small-scale orchard as a cover crop. Apply compost to the soil in a fruit orchard, then till it in. Water the ground with a large sprinkler and let the first flush of weeds germinate, then till again to kill them. Scatter wildflower seeds. Make sure the soil stays moist until seeds have germinated and plants are growing. After the plants finish flowering and have dried, mow or use a weed-whacker on them. This initial seeding should last for five to six years. When weeds gradually dominate the wildflowers, till the ground up and repeat the process. If you do need to till, we suggest you till only half of any orchard in a given year. Tilling can kill the ground nesting bees, so by tilling every other row in each year, you ensure the survival of at least half of your ground nesting bees.

CHAPTER 4

bee garden basics

Once you have identified the bee-friendly plants that are best for your garden, it's time to think about how to put them together in ways that are best for bees, and how to incorporate other elements that bees need. This chapter details what bees require in a garden, such as adequate numbers of flowers blooming seasonally, sufficient patch sizes, sunny spaces, water, and nesting sites. To make your garden safe for bees (and safer for you, too!), we'll explain methods to develop healthy soils and plants that are pest free and disease free.

A Profusion of Blooms Throughout the Seasons

Native bees are active across the flowering season. Specific species of bees emerge from early spring to fall, timed to coincide with the blooming of the native plant species that constitute their preferred pollen sources. The main seasons of bee emergence are late winter/early spring; early spring/summer; summer and late summer/early fall.

Orchard mason bees emerge early in the year, making them effective pollinators of early-blooming orchard crops, such as plums. Some bumblebee species emerge in late winter, as early as January or February in some warmer climates, and other species establish nests later in spring, so you will see more and more species in your garden as the season progresses. Their large size, dark coloration, and fur enable them to tolerate cool or wet conditions better than smaller bees. They also have the ability to detach their flight muscles from their wings and vibrate them to generate heat without flight. This allows them to warm their bodies up enough to function at cooler temperatures than other species.

Previous spread: The repeated plants and profusion of blooms throughout the seasons make this an excellent bee-friendly garden.

Facing page: It is easy to have a variety of bee-friendly plants blooming throughout the seasons by combing early and late blooming plants together.

Honeybees are active during the seasons of the year warm enough for them to fly—generally when temperatures are above 45° to 50°F up to around 100°F. Weather permitting, honeybees are often out early in the spring collecting nectar and pollen from trees like willows, maples, hawthorn, mesquite, and tulip trees (*Liriodendron tulipifera*), and pollen from alders, poplars, and hazelnut. Early-blooming shrubs are important, too, such as manzanita, barberry, wild lilac, redbuds, *Prunus*, willow, and many others.

In warmer areas of the country such as the Southwest or Southeast, honeybees may fly almost year-round. Honeybees expend energy flying on warm days in winter, so in these areas it is important to consider plants that flower in winter. In Florida and the South, wild pennyroyal (*Piloblephis rigida*), red maple (*Acer rubrum*), and black titi (*Cliftonia monophylla*) are winter and early spring bloomers. In the Southwest, a number of plants bloom in low desert areas in the winter: acacia, mesquite, desert marigolds, California poppies, coreopsis, some lupines, some penstemons, desert mallows, some salvias, blanket flowers, and phacelias.

In warm places like the Southwest, bees may be active year round or nearly so, and flowering plants should always be available to them. In this Tucson, Arizona garden, a variety of shrubs and cacti provide pollen and nectar for much of the year.

As bees emerge and forage seasonally, you will see different bees in different seasons in your garden. Having a variety of plants blooming at different times of the year supports many different bees and provides you with a changing landscape of flowers and bees.

To make the process easy, begin with one plant that really appeals to you and will thrive in your area. Choose the next plants based on what will bloom in the same season with this plant and that have colors and form you find pleasing combined with it. For a bee garden, you will want to try and have at least ten different plant species blooming in the same season, but as many as you can manage will benefit bees. Then assemble other groups of plants that bloom either earlier or later than the first group you have chosen, until you have extended flower bloom across as many seasons as your climate allows.

A mix of trees, shrubs, perennials, and annuals, or a combination of any of these elements is sure to give enjoyment and pleasure to you and the bees all season. Trees bloom for two to four weeks, while many shrubs bloom for three to six weeks. A few bloom longer, or bloom in spring and again in the fall. Perennials really perform in terms of flower production during spring, summer, and fall and offer an inviting display of floral possibilities that can fill in the gaps between or after bloom time of shrubs, or in combination with them. Superbloomers (see pages 78–84) that bloom for months without much work on the gardener's part are really good value for your time and resource investment, and obviously are great for the bees. Annuals, with their sheer volume of bloom in season—summer or winter—are fulfilling garden additions for those willing to spend a little more effort to grow them. The more area you can get under constant and varied bloom, the better for the bees and the more beautiful your garden will be.

Goldenrod (Solidago) blooms midsummer through fall, and calamint (Clinopodium nepeta) blooms midsummer through early winter, which are important times for bees who need to gather pollen and nectar in preparation for winter.

Varieties of bee-friendly plants are repeated in close proximity in the garden, creating an impressionistic composition of blue, purple, and white.

Patch Size and Repeated Plants

Although most bees need a variety of pollen and nectar for adequate nutrition, all bees practice "flower constancy," that is, they only visit one species per foraging trip. To make foraging on a plant worthwhile, there has to be a sufficient area of bloom. Research by the Urban Bee Lab at the University of California, Berkeley, has found that bees readily visit patches that are at least 3 by 3 feet (1 square meter) in size.

In terms of designing our gardens, this means that a garden where there is one plant of each species, though they may be plants that are attractive to bees, will likely not be as supportive as a garden that has larger groupings of the same plants. Bees may not visit the single specimens because the plants may not have enough floral rewards to make it worthwhile, or the bee may not be able to find them.

Structuring a garden so that it is bee friendly means that you will want to plant adequate patches of the same flowering plant, or repeat the same plant throughout the garden so that bees can easily forage among them. Having larger patches of the same bloom at one time is easy to accomplish. One tree or large shrub can have many square feet of bloom area. Some individual perennial plants encompass an area of 3 by 3 feet, or two or three can be planted together or near each other to create an adequately sized floral area. The same planting structure can be used with annuals. In formal gardens, a defined area, block of, or edging of the same plant along a path or border can create an effective and striking design composition and provide much fodder for bees. If a more naturalistic scene is a goal, repeating a plant over and over combined with others in a configuration such as would occur in a meadow will create a soft and flowing composition that provides sufficient repetition for bees to be attracted to the resource.

How Many Species?

Bees benefit from a variety of pollen and nectar to feed on, so the greater the diversity of flowering plants in your yard, the better. Honeybees that have only a single source of floral resources at a time suffer nutritional stress. The more species of bee plants you have blooming at one time, the more likely you are to attract and support a large number of native bee species, though whatever you are able to provide is beneficial to them. In a medium size or large garden, it is fairly easy to have at least twenty different plants. Growing a variety of plants enables many possibilities for creative color combinations while pleasing a diversity of bees.

Color, Shape, and Size

When thinking about your bee garden, you will probably want to consider flower color. There is no right group of colors; choices are personal, and while some prefer bright colors, others may be attracted to a quieter symphony of hues. Some find pink and yellow discordant, while others embrace it. At a lecture by two famous garden writers and designers, Rosemary Verey and Christopher Lloyd, Verey gave a spirited and decisive speech on how colors like pink and yellow do not belong together, followed by Lloyd showing slide after gorgeous slide of his garden where pink and yellow occupied the same spaces. They both laughed. Who was right? No one.

You may want to consider the context of the garden when considering color. Under bright skies and harsh light, pale colors can sometimes look artificial. Bright colors such as yellows, oranges, and reds stand up the bright light and often seem to glow in the evening after the sun has gone down. In cooler, moist climates with rich green trees in the view, blues, purples, magentas, pale yellow, and white seem appropriate. To make these paler colors "pop," you can add judicious amounts of yellow or orange flowers such as sunflowers, rudbeckias, or mulleins. A few good annual bright-hued flowers are Iceland poppies, California poppies, peony poppies, and Klondike cosmos. Bright pink flowers like the pink cosmos also serve to brighten quiet compositions. Generally, those colors near to each other on the color wheel are complementary, while strong primary colors opposite each other and are considered contrasting. Too many primary colors can be overpowering and can take over a color composition. Many people tend to sprinkle these colors in their gardens rather than use them on a large scale. It is fun to experiment and change the garden from year to year. Making notes during the season is useful so choice combinations are not forgotten.

The shape and size of flowers also makes a big difference in how they integrate with or dominate a garden, not just for the bees they will attract. Plants like black-eyed susans (*Rudbeckia* spp.) with double blooms tend to have so many flowers blooming together they form almost a solid blob of color and are difficult to integrate into a garden.

*Facing page: The large patches of tall purple verbena (*Verbena bonariensis*) and pink asters make a dramatic landscape statement that's ideal for bees. The scarlet lobelia is a hummingbird plant and also attracts various swallowtail butterflies.*

For bees, these doubled flowers may not allow access to pollen and nectar, or the flowers may produce far fewer floral rewards. Purple coneflower, another daisy with large flowers, has narrow petals and softer colors and so is impossible to overdo, provided you only use the single-flower-form varieties. Single dahlias have similar-size blooms and are spread out over a substantial bush, so are not overwhelming and are easily placed in mixed plantings. Those plants that have more feathery and delicate flowers like asters, joe-pye weed, perennial sunflowers, ironweed, knotweeds, woodland sage, columbine, verbena, or penstemons are easy to weave into perennial borders or are equally effective in a bed by themselves. Smaller "frothers," such as catmint, geraniums, calamint, and oreganos, are hard to overdo because their flowers are subtle. Cupheas have multitudes of narrow yet brightly colored flowers that never overwhelm a color theme. Often cultivars or hybrids of native plants are good choices because their flowers may be slightly larger, more colorful, may bloom longer, or be less lax or sparse in form or habit than the species, but sometimes they can produce less nectar or pollen. Beware of doubled or "dwarfed" annuals or perennials; they are "blobbers" and far less robust.

Shade Versus Sun

Bees need sunny spaces. Most require a certain amount of warmth to fly, so choose a space that receives sun for most of the day for your bee garden. It is often difficult to grow annuals and many perennials under some large tree species such as maples, mulberries, and willows due to shading, but also because of tree root competition. These particular trees have thick and fibrous roots that are greedy for water and nutrients. Willows and maples are very good sources of nectar in early spring and spring, so if you have these trees, they are worth keeping—just don't plant delicate plants under them. Tough ground covers or low shrubs are more suitable for these areas.

Facing page: Coral bells (Huechera spp.) thrive in dry, shady conditions such as those found under some shrubs and trees. Their flowers attract bees and hummingbirds.

Each side of a house, garage, or structure offers great possibilities for small microclimates to suit a variety of plants with specific light and exposure needs. Most houses have roughly north, south, east, and west sides. The north side is often shady at least part of the day and may offer opportunities for shade-loving species that would otherwise be unhappy with direct sun. The south and west sides are generally hot and will get afternoon sun—a boon in a cool, wet climate. The east side of a house can be used beneficially for some plants in hot climates because the combination of morning sun and afternoon shade creates perfect growing conditions for them. The eaves of houses or canopies of trees create shelter from the extremes of cold and sun, like a greenhouse does.

Nesting Sites

Nesting sites are crucial for native bee reproduction. About 70 percent of bees are ground nesting, while 30 percent are cavity nesting. Most native bees are solitary, not social like honeybees, and may nest singly or in small aggregations—numerous bees living in the same vicinity, but not working together like honeybees. Some bee species are very small, and their foraging range limited, so having appropriate areas for nesting close by is helpful. The less distance any bee has to forage from its nest is better. Honeybees literally wear out their wings from flying large distances between their foraging areas and hive. Bees prefer to nest in sunny spots, so position your potential nesting sites in areas that receive sun most of the day. In extremely hot climates, morning sun is best.

Ground Nesting

It is interesting to identify bee-nesting sites. The nests often resemble messy ants' nests with soil pushed up around an entrance, or they can be just a hole in the soil. Sometimes there are a few nests; in other areas there will be many together. In your garden, there are likely to be

Ground nesting bees often dig their own nests, which sometimes, but not always, leaves a mound of soil surrounding the entrance.

Cavity-nesting bees usually nest in preformed cavities like this hole in the wall, or in fence posts, dead trees, cement walls, or adobe.

small groups of bee nests here and there in suitable areas. If you watch a bee nest hole, a female bee will likely enter with heavy saddlebags of orange, yellow, blue, or white pollen, and disappear into the dark confines of the hole. It is hard to imagine small bees having the strength to burrow into soil as much as 2 to 3 feet deep, but some do.

To provide nesting sites for ground-nesting bees in your garden, you need to be sure to leave bare ground and do not till the soil. Sandy, loamy, well-draining soils are best, where water does not stand during rainy weather. The bare areas do not have to be in the middle of your flower bed; they can be along a walk or near a building or even in an area that is not highly visible. Because bees are mobile, their nesting and feeding sites do not have to be right next to each other, but close proximity is better than farther away, especially for the small bees.

Paving and mulching are common in many gardens and mulching is an excellent way to create a healthy soil without tilling and to suppress weeds and retain soil moisture. Both paving and mulching are a necessary part of most gardens, but a bee garden will ideally have areas that have some bare soil because bees can have trouble digging through mulch to nest.

Landscape fabric or plastic should never be used in gardens for weed control because it blocks bee's access to the soil. It also does not allow earthworms and other soil-dwelling organisms involved in decomposition to cycle leaves and other organic matter into soils.

Cavity Nesting

For the 30 percent of bees that are cavity nesting, a garden should have materials available that cavity nesters can use. These include hollow plant stems such as elderberry, old raspberry stems, teasel, or cup plant stalks. Many leafcutter bees nest in plant stems. Some bees like carpenter bees actively drill nest holes in soft wood including unpainted arbors, fence posts, some palm trunks, or dead trees and branches.

Other, smaller carpenter bees in the genus *Ceratina* will use existing small holes in fence posts, old nail holes, or beetle holes for nesting.

Cavity nesters reuse already existing cavities and so are happy to use ones made by humans, although it is better if you can also leave natural materials like logs or hollow plant stems around. Many cavity-nesting bees can be accommodated by putting up predrilled bee block houses, or making bundles of hollow stem plants from the plants mentioned above or from pieces of bamboo. These are easy to make. Beautiful bee hotels made from many different materials, with many variously sized holes, can also be purchased and are beautiful decorative assets in a garden.

Nest blocks can be made from a variety of materials. Combining different textures makes these into wonderful living garden sculptures.

BUILDING NESTING BLOCKS

Nesting blocks can be made by drilling holes between 3⁄32 inch and 3⁄8 inch in diameter, at approximately 3⁄4-inch intervals, in a block of preservative-free lumber. A variety of hole sizes will attract different bee species. You can use anything from leftover 4 by 4 lumber to logs from a fallen tree as the base. In dry areas, adobe blocks can be drilled instead of wood.

The drilled holes should be smooth inside, and closed at one end. Bees create several cells within each drilled hole, so the length is important. Bees prefer holes that are at least 5 inches deep. The height of the block is not as important as the depth, but 8 inches or more is good. Once the nest is built, you can fix the nest to a firm support.

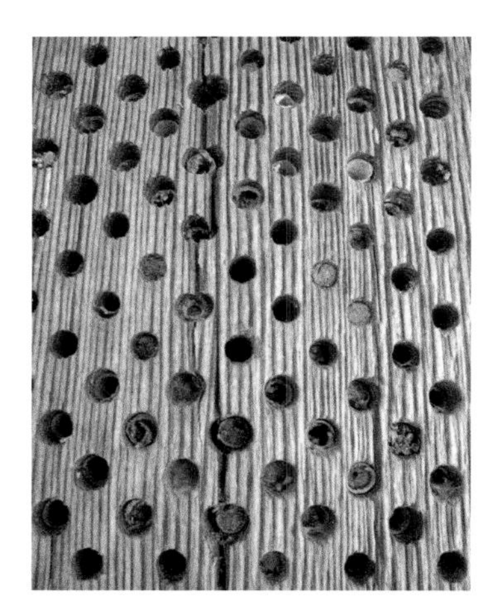

Cavity-nesting bees cap the ends of their nests with materials such as mud (as in this case), leaves, or even flower petals.

Nest sites should be sunny but sheltered from wind and rain. Placing a small roof over the top to keep rain out is beneficial. In hot climates, nests should have protection from direct afternoon sun because nests can overheat. Facing blocks into the morning sun will help nesting bees warm up for flight earlier in the day.

Making nest blocks is a perfect rainy day activity. One winter day we churned out about fifty of them in a short time with various size holes,

Bundling hollow bamboo stems together provides a nest for stem-nesting bees.

all made of scrap lumber. We put them up around a six-acre organic winery garden and surrounding vineyards, expecting to see just a few holes filled here and there. Not long after, there were a variety of bees going in the holes, busy provisioning them with pollen. By the very first summer, every hole was filled and plugged with a variety of plug materials: mud or chewed plant or wood material. Who would think some scrap lumber would generate such interest?

BUILDING STEM OR TUBE BUNDLES

Some plants, like teasel, bamboo, cane fruits, and reeds have naturally hollow stems. Cut the stems into 6- to 8-inch lengths. With a plant that has a segmented stem such as bamboo, teasel, or cup plant, cut the stem just below a node so the closed segment forms one end. If you are using paper drinking straws, fold over one end and staple it. Tie fifteen to twenty pieces into a bundle, with the closed ends together. The bundle can then be secured to a tree, fence, or any structure at any height, as long as the bundle is fairly level. It helps to put the bundle in a protected spot, facing the morning sun. Stem-nesting bees will happily colonize these bundles.

Bundles and blocks should be discarded and replaced after a couple years to avoid a build up of pests and diseases.

BUILDING A BUMBLEBEE BOX

There is scant evidence that bumblebees use nest boxes. Even bumblebee biologists struggle to build nest boxes that will attract bumblebees. The ones that do work tend to have some specific characterisitics. You can find or build a simple wooden box, about 7 inches on each side, made from preservative-free lumber. Drill a few ventilation holes near the top (covered with a screen to deter ants) and some drainage holes in the bottom. Make an entrance tunnel from ¾-inch plastic pipe, marked on the outside with a contrasting color, and fill the box with soft bedding material (for example, upholsterer's cotton). The box must be weather tight; the larvae may become cold in a damp nest, and mold and fungus will grow.

Place the box in an undisturbed site, where there is no risk of flooding. The box should be on or just under the ground. If you bury it, extend the entrance tube so it gently slopes to the surface. Put your nesting box out before you first notice bumblebees in the spring. There is no guarantee that bees will use your box; only about one in four bumblebee boxes is occupied. If it has no inhabitants by late July, put the nesting box into storage until next spring.

Eliminate Pesticides and Herbicides

Pesticides and herbicides negatively affect or kill bees at all different life stages and should never be used in a bee garden. However, this doesn't mean that your garden will be crawing with pests! A variety of factors contribute to creating a healthy, thriving garden. These factors include attracting beneficial insects with appropriate plants, using locally adapted plants, and enriching the soil with compost. Creating nesting habitat for birds also contributes to healthy gardens because most feed insects to their young. This approach works much better for you in the long run, because using pesticides kills off all insects (pests, bees, and beneficial insects), but beneficial insect populations recover much more slowly than pests. This gives pests the advantage and keeps you locked into a detrimental cycle.

Right Plant, Right Place

Selecting plants that are adapted to your region and your particular soils, climate, and pest and disease pressures is essential because it diminishes or eliminates the need for using pesticides. Worldwide, plants are adapted to particular soils and climates. Adapted plants will thrive in your garden with a minimum of effort. Your time can be spent enjoying the flowers and all the bees visiting them, instead of

*Facing page: The showy, yellow coneflower (*Ratibida pinnata*) is native to most states east of the Rocky Mountains and also grows well in the Pacific Northwest. When native and nonnative plants are grown in regions and soils that they are adapted to, they will not be prone to pest and disease problems. Problem plants should be removed and replaced with a species or cultivar that is better adapted.*

fretting over depredations of foliage and flowers from pest insects or diseases. Some plants have evolved in heavy, fertile clay soils, others in gravelly or drier conditions. Some are from the mountains, prairies, meadows, rocky slopes, deserts, chaparral, and others from woodlands, woodland edges, or wetlands. Matching the plants to your particular site conditions helps to ensure you will have a healthy, thriving garden. Plan on grouping plants with the same needs so that they can be easily cared for together.

A plant that has evolved to grow in cool, humid conditions, such as a Japanese maple, will, in a desert environment, be subject to burning on the leaf margins from water often high in salts and hot, drying winds, and will be prone to pests such as spider mites from the lack of humidity and dusty conditions. Plants adapted to dry desert environments and those from cool, moist areas like northern Europe may collapse in regions with hot, very humid summers. Some plants tolerate a huge range of climactic conditions. Many are included in the regional plant lists in the Resources section.

Sometimes there are disease-resistant hybrids or cultivars of plants prone to certain diseases. For example, bee balm is is prone to powdery mildew in some climates; it is a common problem on many of the older varieties but is minimal on some newer hybrids. Adequate watering and removal of any diseased stems also helps keep plants healthy. If powdery mildew is a problem in your area, it is best to research culivars of plants that have resistance to it. Books, the Internet, University Extension offices, and nurseries are good sources of this information.

*Facing page: Catmint (*Nepeta spp.*), bidens (*Bidens ferulifolia*), and tall verbena (*Verbena bonariensis*), bloom for many months if planted in healthy soil.*

CREATING HEALTHY SOIL

The easiest way to create healthy soils is simply to incorporate compost when planting, whether you are planting one plant or preparing a whole area. When beginning a garden, a one-time application can jumpstart your soil. If your soil is clay or sand, put 4 to 6 inches of compost on top of the soil, then dig it in when soil is moist but not wet. If the soil is loam, 3 to 4 inches should suffice. After plants are planted, spread 1 inch of compost around the plants each year to maintain soil fertility.

Water Elements

On warm summer days, honeybees are often seen lined up at edges of fountains, ponds, ditches, and bird baths. Long considered a plague around swimming pools, unfortunate honeybees drown in them during quests for needed summer water. In summer, a colony may need a liter or more a day. Water is essential for cooling the interior of the colony by evaporation. It is also used to dissolve crystallized honey and to dilute honey. Safe water sources for honeybees are simple to make. Floating pieces of corkboard in water creates safe "rafts" for bees to access the water without drowning. Corkboard can often be found at craft stores. Porous wood can also be floated or propped into a water source so bees can access water safely. You can float the corkboard or wood in anything from a plastic bucket to a fancy, decorative fountain.

Water can be used as a focus in the garden. It can be as simple as an attractive birdbath placed at the end of a walk or placed as a focal point, or as elaborate as a pond. Water for honeybees can be as easy as a bowl or saucer filled with decorative rocks and water so that the rocks stick out of the water. Whatever you use, just make sure it is kept clean and regularly refilled.

A garden feature like a birdbath or fountain can be set off by low plants to surround it but not hide the structure. Plantings can help connect these features to the garden and can highlight the placement, or the birdbath can stand alone at the end of a path or in an alcove in a bed. Place water features so they can be easily seen and accessed because much activity happens around them, and you will need to fill

Facing page: Floating corkboard in water gives bees a safe place to drink.

or clean them regularly. Some good low plants to place at the base of fountains are smaller native penstemons, columbine, thyme, smaller pincushion flowers, oreganos, creeping marjoram, *Knautia*, or calamint.

Native bees do not use water for the same purposes as honeybees. It is not used to cool nests or to dilute honey, but a few bee species use mud in nest construction. Some of the ground-nesting and cavity-nesting bees like mason bees use mud to create partitions in nests between larvae. A dripping faucet over bare ground, or a damp bank can be all it takes to supply these bee's needs. You do not need to make mud or dig up soil.

CHAPTER 5

designing your bee garden

Bee-friendly gardens can be composed to fit any yard size—large or small—aesthetic needs, climate, and maintenance requirements. Certain considerations should be taken into account when designing your front yard versus your backyard, as well as the all-important climate considerations of your region. Whether you are a dedicated gardener or someone with little time to spend gardening, you can have a thriving and attractive bee garden.

Designing Your Spaces

There are almost no rules about how to structure a bee garden, but any garden design will need to consider both the functional and aesthetic aspects of outdoor spaces. Many people begin with the essential elements of flow: paths into and through the yard both front and back and places to gather, like seating areas, bench sites, and patios. Water features and large elements like swimming pools also need consideration. Garden uses such as vegetable gardens, garden sheds, and other functions that need specific siting are important to identify. The view from the indoors out can make a big difference in the atmosphere of the house. Once all of these functions and necessities of outdoor spaces are determined, planted spaces can then be arranged around them.

Previous spread: The bee-friendly plants in this garden are grouped in island beds, a very bee-friendly garden design.

Facing page: Bright-hued perennials and annuals surround a fountain with corkboard floating in it. Willow chairs provide a good vantage point to observe bees at work.

Problematic Garden Designs and Solutions

A striking walkway of ancient, gnarled olives, yew, or Italian cypress standing at attention like soldiers makes a strong design statement, but offer little attraction of nectar or pollen to bees. Most wind-pollinated trees like conifers are not bee friendly (as well as the plants listed on page 99). Large-scale use of these plants creates areas that are virtually moonscapes for life. Some Japanese gardens with a preponderance of pines, a glowing carpet of moss or gravel, and scattered evergreen azaleas are not bee friendly, though Japanese maples are when in bloom. Other designs that are typically not optimal for bees are meadow gardens composed of nonnative grasses, modern minimalist gardens, and the typical American suburban yard. However, there are ways to adapt each of these styles for bees.

MEADOW GARDEN

Meadow gardens are very popular. Many are composed of nonnative grasses from Japan, North Africa, Europe, and other places around

the world. Meadows that are strictly made up of these plants are not bee-friendly because there are no flowers for bees to visit, although some of our native grasses supply bees with pollen. Meadows that contain at least 50 percent flowering plants are bee friendly and can be designed with bright or pale colors, and to be tall, medium, or low growing. Lauren Springer Ogen and Scott Ogden's book, *Plant-Driven Design*, has examples of beautiful lower stature meadows with many bee-friendly plants.

MODERN MINIMALIST GARDEN

Some contemporary garden design focuses on a fixed geometry of walls, repeated lines, expansive paving and lawn, blocks of evergreen ground covers, and a few architectural plants. This type of design caters to our sense of order and need for control. Too often these gardens are focused on the word bees are sure to dread: hardscape. Hardscape refers to the nonplant elements in a garden, such as patios and paths. While all gardens require some hardscape, in modern minimalist gardens, plants are only an afterthought to decorate the static pavement, walls, and pebbles that comprise the main garden elements. The ratio of plants to hardscape is very small. As a whole, these gardens are not friendly to bees or other organisms.

What these gardens miss most is the intrinsic connection with the biodiversity of nature. The geometry and form do not engage with the arena of life that surrounds us at all times. Plants are living things, each with a unique physiological method of interacting with the world, each with a distinct cycle of growth and reproduction, and with an inherent beauty generated from these factors. When we walk through gardens filled with living things, we experience this on deeper emotional levels, rather than just on the aesthetic plane.

Facing page: Meadows of ornamental grasses are a popular landscape feature, but most exotic grasses provide no habitat for wildlife. By incorporating at least 50 percent flowering plants, you can provide pollen and nectar for bees and create a soft, colorful scene.

How can a modern minimalist design and a desire for serenity, tranquility, and restraint become bee-friendly? Instead of, say, a single geometric line of olive trees along a low stone wall, simply choose a tree or shrub species that supports bees and has subtle flowers so as to not distract or offend. Trees like maple, tupelo (*Nyssa sylvatica*), basswood or linden (*Tilia cordata* or *T. americana*), Chinese tallow (*Sapium sebiferum*), and the tulip tree (*Liriodendron tulipifera*) have either chartreuse or pale yellow bee-friendly flowers that are not showy. In the shrub department, many hollies, buckthorns (*Rhamnus*), and sweetspire (*Itea*) bushes have chartreuse or pale yellow small flowers attractive to bees over a long bloom period. Myrtle (*Myrtus communis*) has small and delicate white flowers semihidden in handsome, glossy green foliage.

TYPICAL AMERICAN GARDENS

A typical suburban picture that is sadly, from the eyes of a bee, repeated over and over again, is the gravel garden underlaid with plastic weed barrier, and the lawn with a strictly clipped hedge. Bees across the nation suffer from malnutrition from these pollenless and nectarless compositions and a lack of nesting sites.

Hedges that get trimmed without regard to flowering have their floral resources continuously removed; some common boxwood cultivars, yews, and other evergreen shrubs effectively don't flower. But bee-friendly hedges are possible, provided the right species are chosen and trimming occurs after the bloom season and not during it (see page 158).

Most people desire gardens that are easy to care for and maintain, and too often these desires result in lawns or gravel. If one considers the time and resources it takes to maintain a lawn, or to pick weeds from gravel over a year, maintenance on a bee-friendly garden won't occupy more time and certainly gives more pleasure on many levels.

If you desire a simple, easy-care yard, there are bee-friendly answers. A ground cover of creeping rosemary, cotoneaster, wild lilac or New Jersey tea (*Ceanothus*), manzanita, mahonia, sedums, daleas, or

Facing page: Lawns and strictly trimmed hedges create a traditional and tidy scene, but provide no pollen or nectar for bees or other life.

similar plants fit into the almost no-maintenance, single-plant species theme. Better yet, two to three of these species could be combined for an extended bloom season.

As a bee-friendly statement plant, the strawberry tree (*Arbutus unedo*) or flowering crab apples can be used. In the Southwest, a palo verde tree (*Cercidium*) and in the East, one of the sumacs (*Rhus*) can be added. These plants are low-maintenance shrubs or small trees with a strong architectural element and have flowers bees eagerly visit and birds relish.

Another simple solution to maintenance is to simply plant a grove of redbuds or the 'Natchez' crape myrtle or other large shrubs with

distinctive foliage and form in your yard—front or back. Crape myrtles have smooth, distinctive bark that is a feature in itself. They produce pollen, not nectar. Eastern redbud (*Cercis canadensis*) is a deciduous shrub with elegant form and leaves and beautiful magenta flowers in spring. In the Southwest, a number of cactus have wonderful distinctive forms and colorful bee-friendly flowers. Prickly pear cactus (*Opuntia* spp.), barrel cactus (*Ferocactus* sp.), and saguaro (*Carnegiea*) are all very low maintenance garden subjects. They can be interspersed with grasslike sotol or Bigelow's nolina (*Nolina bigelovii*) or the palo verde (*Cercidium*).

Trees may be the simplest easy-care solution yet, and some do well in lawns so that you may have a soft, green lawn and bee fodder as well. Trees like the American basswood, the tulip tree, magnolia, maple, and catalpa all grow in lawn and will shade the yard and feed the

In the Southwest, desert gardens are often sparsely planted but they can still contain many plants friendly to bees, such as prickly pear cactus, which comes in a variety of sizes and flower colors and is very easy to grow.

This small garden in Denver, Colorado, is composed of colorful and bee-friendly plants that bloom over a long season, are low maintenance, and thrive in the arid climate.

bees. Trees like madrone (*Arbutus menziesii*), acacia, palo verde, and velvet mesquite (*Prosopis*) are best with just mulch or compatible shrubs or perennials underneath, and not placed in lawns. Native trees are the best choices for gardens because they cater to many organisms besides bees and are most adapted to conditions found in each area.

If you are thinking of replacing a front lawn with bee-friendly plants and still desire something fairly low growing with minimal maintenance, various creeping thymes, some oreganos, prostate rosemary, some flowering geraniums, turkey tanglefoot (*Phyla nodiflora*), some cotoneasters, mahonia, some wild lilacs (*Ceanothus*), manzanitas, ajuga, Canada anemone (*Anemone canadensis*), some hardy geraniums, low-growing sedums, or similar low plants that form a ground cover are effective visually and contain large amounts of bee fodder. In the Southwest, *Dalea bicolor*, *D. formosa*, and *D. greggii* are colorful and bee-friendly ground covers. Clovers are also highly attractive to bees. Red, strawberry, and white Dutch clover are common, easy to grow perennial varieties. Red clover is mostly visited by bumblebees. All clovers are inexpensive and easy to establish from seed.

LAWN REPLACEMENT

The front yard of this small house on a small lot belonging to Kate Frey's mother had a lawn with an English walnut in it. Once a week, the lawn required mowing, a tedious task. Kate's mother never used the front yard, preferring to sit on the back patio instead. We decided to remove the lawn and put in a bee-friendly garden.

First, cardboard was put over the lawn to kill it, a process referred to as sheet mulching. After putting down the cardboard, the sprayers were converted to a drip system by laying a grid of drip pipe over the top of the cardboard. Four to six inches of compost were put on top of the cardboard to hold it down, increase soil fertility, and hide the dripline. The plants were laid out, a one gallon per hour (1-GPH) dripper was put where each plant was, and the system turned on. Wetting the cardboard and soil underneath makes them both soft and easy to plant into. The cardboard breaks down and is eaten by earthworms when rain or sprinklers moisten it, but it lasts long enough to kill the lawn and any weeds. It is the easiest way to remove a lawn or weeds with very little work while improving soil quality.

A variety of shrubs, perennials, and annuals were planted immediately, right through the sheet mulching, in a cottage garden style. The shrubs used were native barberries, single roses, native sages, wild lilac, and mock orange. The perennials were a variety of smaller native sages, catmint, verbena, irnonweed, sneezeweed, germander, dusty miller, and thyme. For annuals, honeywort and white California poppies were used. What was a nondescript front yard that the neighbors just walked by became a flower-filled place for Kate's mother to meet people and talk—occupations she really enjoyed. People stopped by often and chatted with her and commented on the beauty of the garden, the interesting plants in it, and all the bees, hummingbirds, and butterflies that visited. The garden only took a couple of hours of maintenance a month, about half of what the lawn used to take, and the water bill went down by about one-half as well.

Garden Designs That Are Good for Bees

There are certain garden layouts that are ideal for bees. These include perimeter beds and island beds, which can be incorporated into a wide variety of garden styles and designs. Cottage gardens and flowering meadows are also excellent bee-garden styles. In addition, formal gardens can provide excellent resources for bees provided they contain sufficient plants to support bees.

PERIMETER BEDS

In a small or medium-size garden, you can have planter beds around the perimeter of the space, leaving the interior area for grass, patio, or a surface like decomposed granite. The beds can be made as wide or narrow as the space allows. These perimeter beds can serve to soften fences and create a sense of relaxation and tranquility. Perimeter plantings can connect visually with neighboring trees and shrubs, making the yard appear much larger than it is. Plants can be a mix of tall or small species, as your desires and space dictate. If time to maintain the garden is at a minimum, you may wish to have shrubs dominate with ground covers serving as an underplanting. Shrubs can be chosen to bloom over the course of the spring or summer to offer pollen and nectar all season, with ground covers enhancing the floral abundance. Plants like heal-all, oreganos, thymes, some hardy geraniums, low-growing asters, downy yellow violet, and smooth solomon's seal form attractive ground covers under and around shrubs. Mint can be a good ground cover in a confined space. There are many wonderful varieties including the mouth-watering chocolate mint. The lower-growing prickly pears, smaller daleas, and brittlebush grow well in sunny locations in the Southwest.

*This bee-friendly perimeter bed includes wild buckwheat (*Eriogonum*), native milkweeds, native sages, bluebeard (*Caryopteris*), California fuchsia (*Epilobium*), creeping thymes, rosemary, lavender, culinary sage, Russian sage (*Perovskia*), and a variety of Agastaches in orange and pink. The same plants are repeated over and over, so the long, narrow garden is unified by repeating colors and textures.*

ISLAND BEDS

Rather than encircling the perimeter of a garden, island beds can serve to divide spaces or move people from one part of the garden to another. Dividing up large spaces into smaller ones gives the opportunity to develop different planting styles or color combinations, and for different uses of the garden. If you have enough room, you may

This bee garden is composed of a series of island beds set in decomposed granite and planted with a variety of bee-friendly plants.

want to divide your yard space into a more public area next to the house, with more private sitting areas away from it. Vegetable gardens or orchards can be sited in a different area still. If you like a more formal layout, you can use a series of rectangles to form garden planters and have paths that are straight lines or diagonals to connect them. If you prefer a more naturalistic setting, curving beds and pathways create a relaxed and soft effect. A medium-size shrub such as wild lilac that provides critical early season flowers for bumblebees and can serve to block the total view down a path, giving just a glimpse of what is around the corner, generating a sense of mystery that pulls you down the garden path to explore its potential. Hedges or low fences that delineate garden rooms and have "doors" cut into them can serve the same purpose. We are all drawn to discover what we can just glimpse, and framed views create a sense of desire to know what is beyond. These devices can change a garden from a passive place where all is seen in one view, to an intriguing and exciting environment.

If you have a large lawn and don't want to develop the whole space, marking out potential "islands" with hoses, lines of flour, or white spray paint will allow you to experiment with converting it into a mixed area of bee garden and lawn. The islands allow you to access and view the flowers and bees from a variety of angles—from a broad view to intimate perspectives. Smaller plants can be placed at the edges and larger plants in the middle. A lawn unifies the whole. Because the beds are accessible from every side, maintenance is convenient.

PLANT PLACEMENT WITHIN PLANTING BEDS

For perimeter beds, structuring the plantings so smaller plants are in front and are backed by medium or larger plants allows for good viewing of both bees and plants. The same method is used for island plantings. Shrubs or larger perennials can be used in the middle of a large island planting bed, with smaller plants around the edges of the bed. Some perennials are very large—from 4 to 10 feet tall for some species or varieties—and can substitute for shrubs, though many herbaceous varieties die to the ground in winter. Plants from the prairie regions like the cup plant, some black-eyed susans, ironweed (*Vernonia*), some joe-pye weed (*Eutrochium*), some asters, tall verbena, some sneezeweed (*Helenium*), rattlesnake master (*Eryngium yuccifolium*), and others are striking garden specimens and are easy to grow in much of the nation and make good background plantings or are very effective structurally used in the middle of a large island bed. Many of these plants create a veil-like effect that screens but does not block sight, giving you a hint of what is on the other side. Some large perennials like some perennial sunflowers can spread over time. These plants are best used where there is plenty of room, or against a fence where escape is not possible. They go dormant in winter and are cut down to the ground.

Another method of planting is to use all smaller to medium plants. If you chose plants that have rounded forms such as hardy geraniums, many asters, lavender, gaura, and prairie clover and combine them with plants with strongly upright forms like some of the mulleins, penstemons, foxgloves, or Russian sage, it creates a variety of heights, shapes and forms that create a lively composition. Some smaller plants make good "weavers," like the ornamental oregano cultivars such as 'Santa Cruz', 'Bristol Cross', or to a lesser extent 'Rosenkuppel'. Some coyote mints (*Monardella*) and hardy geraniums are good weavers as well. A newer geranium cultivar called 'Rozanne' has deep blue flowers with surprised white eyes and looks wonderful creeping up the skirts of nearby shrubs.

FORMAL GARDENS

Formal gardens can be wonderful bee gardens. Long swaths of the same plant variety lining a path, fence, or lawn, or forming a planter

Facing page, top: This corner lot has a garden so bright it can be seen from a distance. Plants are placed so small ones are nearer the sidewalk and larger ones near the house, generating a wonderful show, but also privacy for the homeowner.

Facing page, bottom: The tall bee-friendly plants are positioned near the back of this perimeter bed, while low ones in the front soften the edges of a pathway.

bed, create a striking effect. If bee-friendly and long blooming, these plants can supply a tremendous amount of nectar and pollen. For example, a great swath of black-eyed susans (*Rudbeckia hirta* 'Goldstrum') can provide a gleaming patch of bee-friendly yellow daisies. Great swaths of blue catmint (*Nepeta* 'Walker's Low') fronted by a large swath of yellow bidens creates a cheerful scene. Bidens is a tender perennial often grown as an annual, covered with small, yellow daisies for months and requiring no maintenance during the growing season.

Formal gardens are often composed of pathways and planter beds in straight lines. Filling the beds with a profusion of flowering plants creates a scene that is both orderly and dynamic. Doors cut in hedges create a sense of mystery and anticipation that pulls the viewer through.

In the rectangular-shaped mixed perennial border pictured on page 3, an uninterrupted line of deep, delicate blue catmint (*Nepeta* 'Walker's Low'), borders the edge along the long narrow lawn that lies in front of it. The deep, frothing blue acts as a unifying element of the border and sets off many of its colors—while feeding bees. The bright green lawn and the blue catmint are a design element in themselves. Lavender is used in many formal gardens along paths and planters. Even when not in bloom, well-trimmed plants are landscape features.

COTTAGE GARDENS

Many people enjoy a densely planted garden where little soil shows and every inch is maximized and beautiful with flowers that teem with life. The cottage-style garden consists of a variety of intermingling and thickly planted flowering plants—mainly perennials and grasses, but also shrubs that create an impressionistic composition changing with the seasons. The key word is intermingling. Many perennial plants are not at the most attractive as specimens, and are at their best when combined with a group of other complementary or contrasting plants—both of foliage and bloom form and color. This garden type, with its mass of bloom, can be a real visual treat as well as a feast for the bees. Sometimes annuals such as sunflowers, poppies, cosmos, spider flowers, borage, blanket flowers, or wildflowers such as phacelias may be added, giving the opportunity for really vibrant and interesting color combinations and flower form. As many perennials bloom for a portion of the season, the color scheme and potential for bee fodder changes as plants come into and go out of bloom.

Designing Different Parts of Your Garden

No matter what style you choose for your garden, there are design elements to keep in mind for different areas of your garden, such as the front and back yards.

This narrow but profuse front-entry border blocks off a road immediately adjacent to it and creates perfect privacy.

Front Yard

Front yards are almost always underutilized spaces, often composed of unused lawn. Because bee gardens must be flower-filled gardens, transforming a typically generic scene to a gorgeous bee-friendly foreground to your house can enhance both your house and the neighborhood as well. We know of several people who have put in flower-filled, bee gardens and have had the gardens change their whole experience in the neighborhood from one where they hardly spoke with the neighbors, to one where neighbors often stop to admire the flowers and to visit. One friend installed a bee garden in her small front yard after she developed cancer. It became an informal "salon" where she would sit and sip life amid the flowers and wildlife and visit with the many people who stopped by. It became her therapy and ours as well.

In front yards, the entrance to the front door path can be straight, or have gentle curves depending on your taste. Curved paths tend to slow people down, straight paths speed people up. Paths or walks to the front door are important to size adequately as a sense of welcome is generated from sufficiently wide paths, and a lack of welcome is expressed if the path is too narrow. Paths to the front door can be grand or subtle. If the front yard space is large enough, it can be interrupted midway by a small apron of paving at which to pause, and a birdbath, fountain, or bench put there. If the garden is designed as a bee garden, you will certainly want a place to take in the activities going on in it. Benches sited where one can sit will make the garden a place to linger rather than just move through. The area can be divided up into two large beds bisected by a central path or into whatever segments work with the needs or configuration of the space. If there is no path, and the yard is accessed from the driveway, a sitting area or small patio can be created in the middle, with a path leading to the driveway. If you desire privacy, taller plants or shrubs can be planted in a bed next to the sidewalk or against the house for screening purposes without creating a solid wall. Or, if you like more of a ground cover effect, small-stature plants can be used in masses with slightly taller ones placed here and there for interest.

A soft and flowery prairie garden for a front yard could be composed of black-eyed susans, bergamot, purple coneflower, purple prairie clover, smaller anise hyssop cultivars, milkweed, smaller foxgloves, penstemons, prairie phlox, hoary vervain, smaller blazing stars, smaller asters, and the smaller, less aggresive goldenrods. All of these plants are between 1½ and 4 feet tall and grow in the approximately the same conditions. They are native to the Midwest and East Coast but grow well in the Northwest. Many will grow in the Southeast. They are hardy, very showy, flower profusely, and most of them attract butterflies and beneficial insects as well as bees. In the Southwest, plants like desert sage, many penstemons, mealycup sage, blue sage, Cleveland sage, evening primroses, globe mallows, smaller daleas, and encelia could make a colorful flowery meadow adapted to a dry environment.

HEDGES

Privacy is often a concern for front and back yards. Hedges make perfect privacy screens. Many shrubs are low maintenance, and some are very long-lived. Although some hedge plants, such as conifers and nonflowering boxwood, are not bee friendly, many shrubs used for hedging are. Any pruning should happen after bloom. Hedges composed of more than one species are optimal for bees to ensure a greater diversity of floral rewards available to them.

In the dry Southwest, Texas ranger (*Lencophyllum frutescens*) makes a striking silver hedge with pastel-colored blooms. Combined with the red Baja fairy duster, it makes an even more arresting picture. Sugar bush and littleleaf sumac make substantial evergreen hedges in the Southwest. In the South and East, hollies make low-maintenance,

Hedges can offer large areas of bee fodder provided bee-friendly species are used and they are trimmed after the plants have flowered. Pictured is a Japanese privet hedge (Ligustrum japonicum) that has been allowed to bloom.

evergreen hedges that live many years. In the Midwest, summersweet and mock orange are both fragrant and beautiful shrubs. A large, hardy, evergreen shrub that grows well in many areas of the nation is the Carolina laurel cherry.

In the western United States, the California native manzanita, *Arctostaphylos* 'Dr. Hurd', *A. bakeri* 'Louis Edmunds', and *A. densiflora* 'Howard McMinn' are excellent choices, but there are many others. Manzanita flowers are very important for early-emerging bees like those in the genera *Bombus*, *Anthophora*, *Osmia*, *Habropoda*, and *Andrena*. The flowers are a good place to see newly active bumblebee queens. Honeybees and hummingbirds also avidly visit the flowers. *A.* 'Dr. Hurd' is a very drought-resistant manzanita with medium-size, rounded, almost white leaves; it grows to a height of 15 feet. The manzanita *A.* 'Louis Edmunds' grows to a height of 5 to 6 feet and has elegant, sinewy, deep mahogany branches and pink flowers early in the year. The manzanita *A. densiflora* 'Howard McMinn' grows 4 to 5 feet tall.

Strawberry tree (*Arbutus unedo*), a related species from the Mediterranean, also evergreen, is another good choice for an almost no-maintenance hedge. It is very masculine in stance with strongly muscled branches, rough bark, and deep evergreen leaves. It blooms profusely for a long period in fall, with the same heatherlike blooms as the manzanita, and has brilliant red/orange soft fruits in fall that birds avidly feed on. Honeybees, bumblebees, and hummingbirds eagerly visit the blooms. Combining it with other species that thrive in the same conditions will yield bee-friendly blooms from spring through fall. Toyon (*Heteromeles arbutifolia*), native to California, is fairly similar to the strawberry tree, with red berries all winter, and could be added to the hedge, as well as the evergreen coffeeberry and various wild lilacs for prolonged bloom into May, June, and the fall. All these plants support beneficial insects as well. Escallonia, myrtle, and cotoneaster also make good bee-friendly hedges.

BOULEVARD

The boulevard (also called median strip, parkway, tree lawn, parking strip, hell strip, etc.) is the narrow planting area in your front yard between the sidewalk and the street. Boulevards are often neglected because they are detached from the main part of the garden and subject to snow pack, road salt, dog waste, hot temperatures radiated from sidewalks and roads, and trampling by people getting in and out of cars, yet they offer a large area that can be attractively planted, thus enhancing your house and the neighborhood. Plants should be chosen to take some abuse and should be small enough that they won't grow over the sidewalk or into the road. A very attractive boulevard planting

Facing page: The hardy geranium Geranium x cantabrigiense *'Biokovo' creates a delicate sprinkling of white under an Eastern redbud (*Cercis canadensis *'Forest Pansy') that has just finished flowering.*

is creeping woolly thyme, which forms a soft, spreading pool of gray and pink. Another boulevard planting is the hardy, white-flowered geranium (*Geranium* 'Biokovo'), an evergreen ground cover that looks great under an Eastern redbud tree. In the Southwest, a very simple, yet striking boulevard could be composed of acacias, red Baja fairy duster, and yellow golden fleece.

Some of the low-growing sedums such as those used on green roofs are good boulevard plants as well. Some cultivars of goldenrods and asters spread and can be invasive in a small garden, but are completely confined by the cement, as is mint. All can tolerate occasional foot

traffic. Lamb's ears (*Stachys byzantina*) and ajuga are possibilities if you like either silver or purple foliage. Prostrate rosemary (*Rosmarinus officinalis 'Prostratus'*) and the ground cover manzanitas (*Arctostaphylos uva-ursi*) are extremely tough and can take dry, hot conditions as well as some cold. In the Southeast, turkey tanglefoot (*Phyla nodiflora*) makes a tough ground cover many bees love.

Backyard

In the backyard, location of paths, patios, benches, water features, lawns, vegetable gardens, structures, and items to hide or to enhance are best considered before designing planting beds. A backyard garden is like an outdoor room or series of rooms, and the uses and flow of the space must work with your needs. Views are important, too, in some places, and these can be screened with plantings if unsightly or utilized visually to connect to attractive foliage in a neighbor's yard, making the gardens seem connected. Planting beds can easily be designed and placed around these fixed elements.

Although you don't want to overdo the hardscaping in your yard (see page 139), a reasonable amount of paths and patios make a garden more enjoyable and provide access to the garden. Paths generally should be wide enough for two people to walk comfortably side by side. Stepping-stones through beds can provide needed access in a wide border or can lead to a secluded bench or birdbath. Patios can be large enough to accommodate a party, or small enough for just a bench and table. It is important to consider what use each area will have and size each space for its intended use. It is easy to stage a table and chairs in an area being considered for a patio and measure how much space they take up and to move easily around them. Patios, seating areas, and paths can be outlined with hoses, landscape flags, or lines made from sprinkled flour, and tested for practical or aesthetic application from many angles, including visually from inside the house, before any work occurs.

Often a patio or terrace is placed outside the back door so that it is easy to sit or eat outside the house. Planting beds against the house and around the patio can soften the hard feel of the paving. These planting beds can easily be filled with bee friendly plants. The rest of the garden needs to work with the shape and size of your site and your needs and desires.

Facing page: In this backyard bee garden in Seattle, Washington, a carefree mixture of blooms from colorful perennials look almost like butterflies hovering above the foliage.

Placing planting beds on either side of a door or gateway generates a sense of welcome. In this garden (pictured above right), a large planter has been placed in front of and around three sides of the vegetable garden. An old Victorian door painted green forms the entrance next to an old sink for washing vegetables picked from the garden. Brightly colored bee- and hummingbird-friendly shrubs and perennials froth along a wire fence on either side. Old-fashioned roses are included in the mix for a fragrant, romantic feel. In spring, an old-fashioned and pastel mix of roses, manzanita, lamb's ears, red salvias, perennial horehound (*Marrubium*), lavenders, and catmints bloom. In midsummer, bronze fennel, goldenrods, deep raspberry *Knautia*, germander, and *Agastache* 'Purple Haze' flower. In late summer, oreganos, sedum, perennial sunflowers, asters, California fuchsias, sunset agastache (*Agastache rupestris*), and another wave of bloom from the sages creates a sunset-colored haze. In the middle of this abundant, buzzing mix of blooms is the vegetable garden. There are plenty of bees to pollinate the vegetables and fruit trees. Guests are always drawn to the green door and must see what is behind it.

Above, left: Creating planting beds next to a building grounds it in space and generates a sense of welcome and interest.

Above, right: An old door leading to a vegetable garden becomes a place of fun and mystery adorned and surrounded with plants. Visitors are drawn to see what is on the other side.

HIDING FENCES AND WALLS

In the backyard, fencing is often a dominant element. But fences can disappear when large plants are placed in front of them, causing the boundaries of your garden to appear larger. Besides shrubs, some perennials are tall and very showy in summer, though often go dormant and die to the ground in winter. They form lively and colorful screens that are best used where they have room to express themselves. A few of these plants are the cup plant (*Silphium perfoliatum*), a striking, perennial yellow daisy that grows 3 to 10 feet tall and prefers moist soil. Many of the deep pink or magenta joe-pye weeds (*Eutrochium*) and ironweed (*Vernonia*) combine well with it in terms of the same water and soil requirements and tall stature. The cup plant is visited by many bees both large and small, such as green sweat bees, leafcutter bees, long-horned bees, and bumblebees. In general, the open composite flowers, with easily accessible nectar and pollen, cater to many flower visitors, including bees with short and long tongues. Striking, architectural leaves surround the cup plant stem and form a reservoir of water for insects and birds to drink from. Birds like finches and sparrows feed on the seeds.

Below, left: Foxgloves provide a seasonal, attractive and bee-friendly screen in front of a brick wall.

*Below, right: The large stature of the cup plant (*Silphium perfoliatum*) breaks up and softens the imposing feel of this large wall. Wildflowers in complementary colors in front of it create a cohesive scene.*

Garden Design in Different Climates

Planting native plants and plants from climates similar to your region from elsewhere in the world creates gardens that express the particularities of the local region and place through the appearance or scene the plants create. When you are in nature you can tell where you are geographically by the surrounding plants, whereas in gardens, often you have no clue to your location, because gardens are filled with a variety of nonnative plants from many ecoregions of the world. Nonadapted plants frequently need extensive intervention from us in terms of soil amendments, water, and pesticides in order to thrive, so they are not just visually incongruous; they can be environmentally detrimental as well. For example, in desert or arid regions, plants from coastal areas, the Midwest, or eastern United States are often planted. In arid Bakersfield, California, redwood trees from northern coastal regions and lawn grasses from the Midwest dominate much of the commercial landscapes, consuming vast amounts of finite water in order to survive.

Susan Damon's cottage-style Minnesota garden contains a profusion of native plants that thrive in the extremely cold winters and warm, humid summers of the Midwest and attract many bees and other wildlife.

Michael McDowell's Plano prairie garden in Texas is filled with bee-friendly plants rich in color and interesting form.

In contrast, the Tucson, Arizona, landscape is mostly composed of native plants or plants from other similar desert regions. The plants and city architecture together creates a strong identity and very dynamic and atmospheric sense of place that identifies where you are geographically. Using the "right plant, right place approach" helps generate healthy plants and gardens that are not prone to pests and diseases.

Plants from the desert Southwest have very small, delicate leaves, and the plants tend to be wispy and open, generating a light and airy feeling. Plants from the Midwest, East, Northeast, and the higher-rainfall regions of the Northwest generally have larger leaves of deep green that cast more dense shade and have a stronger presence. Perennials can be quite large. In the Rocky Mountain area, in high elevations such as around Denver, rainfall is low and the growing season is short. Native plants can have delicate leaves due to the aridity of the climate, but generally have greener foliage than the Southwest. For bee gardens, soil- and climate-adapted plants are vital because they will be not prone to stress and pests and diseases, and so will not require pesticides that are so detrimental to bees.

Native plants are important because native bees have coevolved with native plants and prefer these plants over nonnative ones. Planting at

Debbie Bosworth's large bee and butterfly border in Plymouth, Massachusetts, brims with bee-friendly perennials and shrubs such as 'Black Knight' butterfly bush, purple coneflowers (Echinacea purpurea), bee balm (Monarda didyma 'Jacob Kline'), catmint (Nepetia 'Giant'), black-eyed susans (Rudbeckia hirta), and 'Lemon Queen' sunflowers.

least 50 percent native plants and then observing the bees that visit them seasonally will allow a glimpse into the inner workings of nature in your area. Bee and bee/plant partnerships can change in a short geographic area if the plant composition and climate changes. Understanding the relationship between plants and their pollinating visitors is to see into another world and helps us form an attachment to it. We cannot look at either the plants or the bees in the same way after the experience of observing them together.

For lists of native plants, please see the Regional Plant Lists on pages 190–206.

In Lauren Springer Ogden's garden in Fort Collins, Colorado, pink asters bring cheer to a fall scene.

*Janet Allen has been bee gardening in upstate New York since 1999. Her pesticide-free suburban yard is filled with native plants, such as joe-pye weed (*Eutrochium maculatum*), purple coneflowers (*Echinacea purpurea*), and blue vervain (*Verbena hastata*), along with nesting sites for a variety of bees. These simple measures create a haven for bees and an enjoyable habitat for her family, and increase fruit and vegetable production in her edible garden.*

Adding Bee-Friendly Plants to an Existing Garden

Many people begin with an existing garden and turn it into a bee garden over time, rather than removing all the plants and starting from scratch. It isn't difficult to successfully incorporate bee-friendly plants into an existing garden.

Spaces in a garden open up when some plants become overmature, die, or simply don't thrive. Sometimes a space opens up when you simply decide a plant does not appeal to you. Each potential open area needs to be evaluated for its mix of sun and shade, irrigation schedule, soil fertility, size of space, and neighboring plants. Then new plants should be chosen accordingly.

New plant selections should fit in with the existing plants' regimen of water and soil fertility. For example, in a border of drought-resistant plants, you will want to add plants that tolerate scant amounts of water; in a garden with plants requiring fertile or moist soils, any new plant should have the same basic requirements. Some species of native plants such as goldenrods, wild buckwheats, prairie blazing star, milkweeds, and wild lilacs (*Ceanothus*) are specific to geographic regions across the country and can be locally sourced. A number of common garden plants, including coreopsis, blanket flower, sneezeweed, sunflowers, catmint, lavender, oreganos, cotoneasters, mock orange, goldenrod, bergamot, redbud, flowering crab apples, locust, and others tolerate a wide range of wet to fairly dry soils and cold and heat—but not extremes in either direction—and are good choices for the beginning gardener.

You will also want to consider flower color and configuration, plant form, foliage type, and bloom time when choosing plants to add to an existing garden. All gardens have an aesthetic style and including plants that fit with or will add to an existing scheme will help give visual cohesion to your garden. Plants from different climate regions have distinctive appearances, and for a cohesive aesthetic theme as well as grouping appropriately for plants' cultural needs, these elements need considering.

Bloom time is important as well. There may be a bloom gap that needs filling. For example, you may have a number of plants that bloom in spring and others that bloom July through October, but April through June may have few blooming plants. Taking the opportunity to fill this gap will help your garden provide more floral resources for bees.

Correct spacing between existing plants is also important. Each empty space has certain dimensions. Make sure to select a plant that won't grow too large for the space and smother its neighbors or shade them out. Some plants have great ability to spread, while others spread extremely slowly or not at all. The bog sage (*Salvia uliginosa*) grows to be about 4 feet tall and spreads over time, depending on the soil conditions. It needs space. Likewise many goldenrods, some perennial sunflowers, select sages like *S. gauranitica* and *S. elegans*, the narrow leaf milkweed (*Asclepias fascicularis*), the ornamental oregano 'Hopleys', culinary oregano, comfrey, and mint spread over time. They are not good candidates for a small empty spot in a garden or small gardens. Good choices for small spaces are well-behaved and delicate plants like calamint (*Clinopodium nepeta*), columbine, purple prairie clover, blanket flower, and sea hollies, but there are many more plants. Some plant genera like *Lupinus* have numerous species to chose from in large and small sizes, and annual and perennial varieties. If you are looking for smaller shrubs, some good ones that come in dwarf sizes as well as large are the strawberry tree, mock orange, as well as some grevilleas, cotoneasters, manzanitas, myrtles, mahonia, escallonia, hydrangeas, spireas, sweetspire, and wild lilacs.

CHAPTER 6

beyond your own backyard—becoming a bee activist

Becoming a bee activist can be easy. There are simple things you can do to help change the landscape to one that both serves humans and helps pollinators. The critical steps are to plant gardens that support biodiversity, identify and protect nesting habitat, contribute to research on bees, and spread the word!

If you've begun a bee garden at home, you've already taken an important first step. As we know, an effective bee-friendly garden doesn't have to be large (see page 60), and yet, home gardens can make a powerful impact (see pages 12–13). Beyond that, we should be striving to enhance biodiversity in all our human-dominated landscapes—from home gardens to corporate headquarters to city parks. Since we cannot create nature reserves everywhere they are needed, taking steps to increase the quality of our managed landscapes as reservoirs of biodiversity will minimize the loss of natural habitats and provide benefits to humans through improved water quality, flood control, enhanced pollination, and greater beauty. Our neighborhood gardens and community gardens can easily be part of this great possibility simply by tailoring planting for pollinators and other wildlife.

Ecologists know there is a strong relationship between the size of an area and the number of species found there. This happens because in a larger area there are more types of habitats and different species can be accommodated in those differing habitats. A large habitat is also likely to be able to support a large population of any given species, which decreases the potential for that particular population to go

Previous spread: The carefree and bee-friendly flowers of Bidens ferulifolia *'Hawaiian Flare' cover the plant all season.*

Facing page: Green sweat bee (Agapostemon texanus).

extinct due to random causes. So, if we can create larger areas for species to use, we help increase biodiversity. We can be thoughtful about our human-dominated areas and take steps to make them useful for different species.

Imagine if we could ensure that every house, school, coffee shop, and municipal building had three things: a portion of their plantings devoted to bee-friendly native plants, some bare ground for ground nesters, and some sites for cavity nests. We could build a chain of pollinator friendly habitats across the United States that would support bees. This chain of resources would enhance what is already provided in nature reserves and gardens and help to build the infrastructure that will maintain larger populations of bees. In addition to supplementing local habitat resources, these sites may provide the critical link that will allow bee populations to respond to climate change. Rather than encountering inhospitable landscapes with no resources that inhibit their ability to move northward as the climate shifts, these enhanced sites could create the corridors necessary to support species as they move or migrate.

Protect and Promote Native Plants

While many plants provide resources for bees, native plants are especially beneficial. These are the plants that have evolved with the local pollinators and evolved in the local habitats. They are likely to support specialist species and be easier to grow without the aid of pesticides and herbicides. So first and foremost, we need to grab whatever opportunities exist for preserving important native wildflower areas. As human populations grow, more and more lands will be under pressure from development. The Nature Conservancy, Trust for Public Land, and the many local and regional land trusts are organizations dedicated to preserving wild lands and they deserve your support. In addition, city, county, state, and national park lands are often under pressure and budget cuts, and welcome any volunteer time, money, and votes for ballot initiatives. Making sure that we preserve those wild areas that support biodiversity is important.

While it's important to plant native plants in your own garden, it is also important to reach out and push for public garden spaces to include native plants. For example, Sarah Burgman has led the creation of a mile-long "Pollinator Pathway" designed to connect the fragmented landscapes across the city of Seattle. To get started in your own area, you can use The Great Sunflower Project's (see page 178) checklist for pollinator-healthy gardens. This tool can be used for evaluating home gardens as well as public spaces such as playgrounds and community gardens. Some people have even brought the list to town council meetings to promote more pollinator-friendly habitats.

Identify and Protect Nesting Habitat

Although well-maintained, tidy landscapes are highly valued aesthetically, bee populations suffer when homeowners and land managers fill in areas of bare ground, remove dead trees and brush, cut down native vegetation, mow or apply herbicides to flowering meadows or roadsides, and clean up every cane in their berry patches. These are the resources many bees use for overwintering. And, it is not just bees that need them. Other insects and birds often use these same resources.

Identifying bee nesting sites is the first step in protecting them, but finding bee nests can be difficult. One of the ways to do it involves the use of clear plastic glasses: When you find a perfectly round hole in the ground, place the glass over it and then wait or come back in fifteen minutes. If a female bee is using that as a nest site, she will be buzzing around inside or outside of the plastic glass. Her presence indicates that the hole is an active nest.

If you're unsure about identifying bee nests, start by looking in your own yard. A small pile of soil that resembles a messy ants' nest can be a clue that it may be a bee nest. The holes are very small—bee size, but you can readily see small bees entering and exiting. Wearing reading glasses or using binoculars can help make the small details of the bees easier to see. Once you recognize what the nests are, you will start seeing more of them other places in the garden and in others' gardens. Familiarity will help with identification.

When you find areas that are being used, point them out to the land manager for that site, and encourage that person to not disturb the areas with digging or irrigation. Often these areas are used by many different species across years because they have the right soil properties for being excellent nest sites.

Citizen Science

Over the past ten years, opportunities have opened for citizens to participate in research about bees. The best of these projects not only contribute to science but also help you learn more about your own yard.

1 **GREAT SUNFLOWER PROJECT** (www://greatsunflower.org). The Great Sunflower Project is the oldest and most well established of the projects. They have three main ways to get involved. The Safe Gardens for Pollinators project uses 'Lemon Queen' sunflowers and focuses on identifying the effects of pesticides on pollinators. The Pollinator Friendly Plants program is working to identify the critical plants for supporting pollinators in different regions. Both of these have participants do timed counts of pollinators visiting flowers and usually only take a few minutes. The third option is the Great Pollinator Habitat Challenge, which is focused on evaluating and improving habitat for pollinators. Participants run through a checklist and identify how green spaces can be enhanced to support pollinators.

2 **BUMBLE BEE WATCH** (www.bumblebeewatch.org). Bumble Bee Watch is a collaborative effort to track and conserve North America's bumblebees. Participants photograph bumblebees and upload them through the website. The data contributes to understanding how distributions of bumblebees are changing and identifying areas and species that need conservation efforts.

3 **BUMBLE BEE CONSERVATION NEST SURVEY** (www.xerces.org/bbnest). Little is known about nesting in bumblebees, which makes it difficult to work on conservation for these bees. To begin to understand whether limited nesting habitat is affecting bumblebees, this Xerces Society project asks that any time you find a bumblebee nest, you fill out a short online survey.

4 **URBAN POLLINATION PROJECT** (www.nwpollination.org). This is an innovative project in Seattle focused on using tomato plants to learn about pollination. Volunteer citizen scientists grow three experimental tomato plants: an open-pollinated control (a regular plant), a self-pollinated plant (covered with a net so bees cannot pollinate flowers), and a plant that receives extra "buzz pollination" with a tuning fork. They then measure the number and size (volume) of tomatoes produced over the season by each plant. By comparing among these groups, they can determine the quality of pollination.

5 **ZOMBEE WATCH** (www.zombeewatch.org). Zombee Watch is focused on identifying where the zombie fly (*Apocephalus borealis*) is parasitizing honeybees and documenting how often honeybees leave their hives at night, even if they are not parasitized by the zombie fly. Participants collect honeybees found under lights in the morning or set up light traps near honeybee hives and report to the website.

Long-horned bee (Melissodes robustor).

Teach Others About Bees

Part of creating a pollinator-friendly world is spreading the word to get others to join in with you. The more habitat that you can improve, the healthier our pollinators will be. Even if you do not think of yourself as an outreach sort of a person, you can do some things to help. Think about what you like to do and how you can use those interests to spread the word! Here are some of our favorite ideas:

- Take a child or school group for a walk in your garden, have them interview you about pollinator habitat, and send their words, photos, or video to the Great Sunflower Project.
- Plan an outing for your community or group and find a pollinator gardener or pollination expert to lead it.
- Get a study group together to learn about pollinators.
- Photograph or draw as many pollinators as possible, using Discover Life's online identification keys (www.DiscoverLife.org) or a pollinator guide.
- Recruit an artist to create a mural or mosaic to celebrate people's connections with pollinators, to build a new water element for a garden, or to jury an exhibition of children's drawings of pollinators at the local library.
- Hold a concert in a garden celebrating the sounds of nature.
- Find a local senior center, school garden, or public garden and do a pollinator habitat assessment and help them develop a plan for implementing improvements the following year.

We hope you will take the information in this book and use it to plant a bee-friendly garden and help others to do so, too. The joy of a garden filled with the myriad and fascinating forms of a diversity of bees visiting the lovely swaths of floral beauty you have created is a pleasure too little realized, but easily accomplished.

Happy bee-friendly gardening!

Resources

Bee-Friendly Nurseries

Southeast

Abide-A-While Garden Center
1460 Highway 17
North Mount Pleasant, SC 29464
843-884-9738
www.abideawhilegardencenter.com

All Native Garden Center & Plant Nursery
300 Center Road
Fort Myers, FL 33907
239-939-9663
www.nolawn.com

ATS Nursery
2514 Vada Road 39817
Bainbridge, GA
229-416-6282

Biophilia Native Nursery
12695 County Road
Elberta, AL 36530
251-987-1200
www.biophilia.net

Carolina Wetland Services
550 E Westinghouse Boulevard
Charlotte, NC 28273
704-527-1177
www.cws-inc.net

Growild, Inc.
7190 Hill Hughes Road
Fairview, TN 37062
615-799-1910
www.growildinc.com

South Central

Bamert Seed Company
1897 County Road 1018
Muleshoe, TX 79347
800-262-9892
www.bamertseed.com

Maypop Hill Nursery & Publications
P.O. Box 123
4979 Spec Garig Road
Norwood, LA 70761
225-629-5379

Natives of Texas
4256 Medina Highway
Kerrville, TX 78028
830-896-2169
www.nativesoftexas.com

Pine Ridge Gardens
P.O. Box 200
London, AR 72847
479-293-4359
www.pineridgegardens.com

Southwest

High Country Gardens
Online and phone orders only
Santa Fe, NM
800-925-9387
www.highcountrygardens.com

Mountain States Wholesale Growers
10020 West Glendale Avenue
Glendale, AZ 85307
800-840-8509
www.mswn.com

Plants of the Southwest
Santa Fe Nursery Location
3095 Agua Fria Road
Santa Fe, NM 87507
505-438-8888

Albuquerque Nursery Location
6680 4th Street NW
Albuquerque NM 87107
505-344-8830
Mail Order
800-788-7333 (seed orders only)
plantsofthesouthwest@gmail.com
www.plantsofthesouthwest.com

Pacific Northwest/West

Annie's Annuals and Perennials
740 Market Avenue
Richmond, CA 94801
888-266-4370
www.anniesannuals.com

City People's Garden Store
2939 East Madison Street
Seattle, WA 98112
206-324-0737
http://citypeoples.com/gardenstore/

Dave Wilson Tree Nursery
Locations across the country
www.davewilson.com

Digging Dog Nursery
31101 Middle Ridge Road
Albion, CA 95410
707-937-1130
www.diggingdog.com

Garden Fever
3433 Northeast 24th Avenue
Portland, OR 97212
503-287-3200
www.gardenfever.com

Mountain Valley Growers Herb Nursery
38325 Pepperweed Road
Squaw Valley, CA 93675
559-338-2775
www.mountainvalleygrowers.com

Rocky Mountain/Intermountain West

High Country Gardens
Online and phone orders only
Denver, CO and Santa Fe, NM
800-925-9387
www.highcountrygardens.com

Sharp Brothers Seed Company
Phone orders only
Healy, KS
800-462-8483
www.sharpseed.com

Midwest

Dropseed Native Nursery
1205 South Buckeye Lane
Goshen, KY 40026
502-439-9033
www.dropseednursery.com

Prairie Frontier
Online and phone orders only
Waukesha, WI
262-544-6708
www.prairiefrontier.com

Prairie Nursery
W7262 Dover Court
Westfield, WI 53964
800-476-9453
www.prairienursery.com

Roundstone Native Seed Company
Online, phone, and mail orders only
9764 Raider Hollow Road
Upton, KY 42784
270-531-3034
www.roundstoneseed.com

Shooting Star Nursery
160 Soards Road
Georgetown, KY 40324
502-867-7979
www.shootingstarnursery.com

Northeast

Earth Tones Native Plant Nursery and Landscapes
212 Grassy Hill Road
Woodbury, CT, 06798
203-263-6626
www.earthtonesnatives.com

Fieldstone Gardens, Inc.
55 Quaker Lane
Vassalboro, ME 04989
207-923-3836
www.fieldstonegardens.com

Found Well Farm
730 Borough Road
Pembroke, NH 03275
603-228-1421
www.foundwellfarm.com

Native Haunts
Online and phone orders only
Alfred, ME
207-604-8655
www.nativehaunts.com

New England Wetland Plants, Inc.
820 West Street
Amherst, MA 01002
413-548-8000
www.newp.com

Pan's Acres Nursery LLC
62 Colburn Road
Canterbury, CT 06331
860-662-0203
www.facebook.com/pages/Pans-Acres-Nursery-LLC/330296982965

Project Native
342 North Plain Road
Housatonic, MA 01236
413-274-3433
www.projectnative.org

River Berry Farm
191 Goose Pond Road
Fairfax, VT 05454
802-849-6853
www.riverberryfarm.com

Mid-Atlantic

Environmental Concern
201 Boundary Lane
St Michaels, MD 21663
410-745-9620
www.wetland.org

Ernst Conservation Seeds
8884 Mercer Pike
Meadville, PA 16335
800-873-3321
www.ernstseed.com

Fiddlehead Creek Native Plant Nursery
7381 State Route 40
Hartford, NY 12838
518-632-5505
www.fiddleheadcreek.com

Native Landscapes & Garden Center
991 NY-922
Pawling, NY 12564
845-855-7050
www.nativelandscaping.net

Pinelands Nursery
323 Island Road
Columbus, NJ 08022
609-291-9486
www.pinelandsnursery.com

Toadshade Wildflower Farm
Online and mail orders only
53 Everittstown Road
Frenchtown, NJ 08825
908-996-7500
www.toadshade.com

Turtle Tree Seed
Online, phone, and mail orders only
10 White Birch Road
Copake, NY 12516
800-930-7009
www.turtletreeseed.org

Bee-Friendly Public Gardens

Arizona-Sonora Desert Museum
2021 North Kinney Road
Tucson, AZ 85743
520-883-1380
www.desertmuseum.org

Denver Botanic Gardens
1007 York Street
Denver, CO 80206
720-865-3501
www.botanicgardens.org

Desert Botanical Garden
1201 N Galvin Parkway
Phoenix, AZ 85008
888-314-9480
www.dbg.org

Longwood Gardens
1001 Longwood Road
Kennett Square, PA 19348
610-388-1000
www.longwoodgardens.org

Mendocino Botanic Garden
18220 CA-1
Fort Bragg, CA 95437
707-964-4352
www.gardenbythesea.org

Naples Botanical Garden
4820 Bayshore Drive
Naples, FL 34112
239-643-7275
www.naplesgarden.org

Natural History Museum of Los Angeles County
900 Exposition Boulevard
Los Angeles, CA 90007
213-763-DINO
www.nhm.org/nature/visit/nature-gardens

Saint Louis Zoo
1 Government Drive
St. Louis, MO 63110
314-781-0900
www.stlzoo.org/conservation/wildcare-institute/center-for-native-pollinator-conservation

Smithsonian Butterfly Garden National Museum of Natural History
9th Consitution Avenue NW
Washington, D.C. 20050
202-633-2220
www.mnh.si.edu/museum/butterfly.html

Tohono Chul Park
7366 N Paseo Del Norte
Tucson, AZ 85704
520-742-6455
www.tohonochulpark.org

University of California Botanical Garden at Berkeley
200 Centenial Way
Berkeley, CA 94720
510-643-2755
www.botanicalgarden.berkeley.edu

University of California at Davis Haagan-Daz Honeybee Haven
1 Bee Biology Road
Davis, CA 95616
hbhinfo@ucdavis.com
http://hhbhgarden.ucdavis.edu

University of Minnesota Landscape Arboretum
3675 Arboretum Drive
Chaska, MN 55318
952-443-1400
www.arboretum.umn.edu

Wave Hill Garden
W 249th Street
Bronx, NY 10471
718-549-3200
www.wavehill.org

Bidens ferulifolia 'Hawaiian Flare'.

Recommended Books

Ashworth, Suzanne. *Seed to Seed: Seed Saving and Growing Techniques for Vegetable Gardeners*. Decorah, IA: Seed Savers Exchange, Inc., 2002.
Beck, Travis. *Principles of Ecological Landscape Design*. Washington, DC: Island Press, 2013.
Buchmann, Stephen L., and Gary Nabhan. *The Forgotten Pollinators*. Washington, DC: Island Press, 1996.
Chambers, Nina, Yajaira Gray, and Steven Buchmann. *Pollinators of the Sonoran Desert*. Tucson, AZ: Arizona-Sonora Desert Museum, 2004.
Frankie, Gordon W., Robbin W. Thorp, Rollin E. Coville, and Barbara Ertter. *California Bees and Blooms*. Berkeley, CA: Heyday Press, 2014.
Griffin, Brian L. *The Orchard Mason Bee*, 2nd ed. Bellingham, WA: Knox Cellars Publishing, 1999.
Grissel, Eric. *Insects and Gardens*. Portland, OR: Timber Press, 2001.
Holm, Heather. *Pollinators of Native Plants*. Minnetonka, MN: Pollination Press, 2014.
LeBuhn, Gretchen. *Field Guide to Common Bees of California*. Berkeley, CA: University of California Press, 2013.
Mielke, Judy. *Native Plants for Southwestern Landscapes*. Austin, TX: University of Texas Press, 1993.
Ogden, Scott and Lauren Springer Ogden. *Plant-Driven Design*. Portland, OR: Timber Press, 2008.
Pellett, Frank C. *American Honey Plants*. Hamilton, IL: Dadant & Sons, 1977.
Proctor, Michael, Peter Yeo, and Andrew Lack. *The Natural History of Pollination*. Portland, OR: Timber Press, 1996.
Tallamy, Douglas W. *Bringing Nature Home*. Portland, OR: Timber Press, 2009.
Whittlesey, John. *The Plant Lover's Guide to Salvias*. Portland, OR: Timber Press. 2014.
Xerces Society. *Attracting Native Pollinators: The Xerces Society Guide*. North Adams, MA: Storey Publishing, 2011.

Pollinator Organizations

The Pollinator Partnership
423 Washington Street, 5th Floor
San Francisco, CA 94111
415-362-1137
www.pollinator.org
A nonprofit focused on promoting the health of pollinators. As the parent organization of NAPPC (North American Pollinator Protection Campaign), they initiated National Pollinator Week, and you can register your pollinator garden on their S.H.A.R.E. Map to take part in the Million Pollinator Garden Challenge (www.millionpollinatorgardens.org).

University of California at Berkeley Urban Bee Lab
www.helpabee.org
Led by Dr. Gordon Frankie, this organization is dedicated to researching and disseminating information about native bees in urban areas.

USDA Pollinating Insects—Biology, Management, and Systematics Research
www.ars.usda.gov/main/site_main.htm?modecode=20-80-05-00
This is the premier USDA group focused on pollinators. Their mission is to encourage and conserve non-Apis (nonhoneybee) bees for sustainable crop production, and they house the world-renowned bee museum, the US National Pollinating Insects Collection in Logan, Utah.

Xerces Society
628 Northeast Broadway, Suite 200
Portland, Oregon 97232
855-232-6639
www.xerces.org
A nonprofit focused on invertebrate conservation. The pollinator conservation arm of this organization promotes conservation on farms, among other things.

Online Information

Attracting Pollinators to Your Garden Using Native Plants
www.fs.fed.us/wildflowers/pollinators/documents/AttractingPollinatorsV5.pdf
An attractive guide to managing a garden for pollinators, produced by the US Forest Service.

BugGuide
www.BugGuide.net
An online photo repository of insect pictures with expert identifications. It is one of the best collections of insect photos in the world.

Discover Life
www.discoverlife.org
A website with online keys for identifying bees and many other plants and animals. Within the keys are excellent species descriptions and maps of distribution.

The Great Sunflower Project
www.greatsunflower.org
A citizen science project focused on pollinators that includes numerous resources on planting, bees, and pollinator conservation.

The Handy Bee Manual
http://bio2.elmira.edu/fieldbio/beemanual.pdf
A guide for scientists and passionate amateurs on the best practices for studying bees.

Insect Visitors of Illinois Wildflowers
www.illinoiswildflowers.info/flower_insects
A database of insects that visit the various wildflowers of Illinois for nectar or pollen.

The Pollination Homepage
www.pollinator.com
A portal to pollination information and images.

Pollinator Partnership
Pollinator Friendly Ecoregional Planting Guides
www.pollinator.org/guides.htm
Guides to flowers that support pollinators designed for different regions of the United States.

University of Florida Extension
Institute of Food and Agricultural Sciences
Beekeeping: Florida Bee Botany. Cir. 686
http://entnemdept.ufl.edu/honeybee/extension/Beekeeping%20-%20Florida%20Bee%20Botany.pdf
A guide for Floridians to the important nectar plants of that state.

Regional Plant Lists

The plants on the following lists were chosen because they appeal to many bees and they are good garden subjects in the correct situation. Note that all information is to the best of our knowledge. Plants can behave differently in different climates and soils. There are more plants besides the ones listed here that are both bee-friendly and good garden specimens.

There is great variation in each region's climate, soils, elevation, and topography. In a local area of only fifty miles, you can find surprising differences in soils, precipitation, and temperature even though the whole area may be in the same climate zone. Even in areas as small as an urban garden, each side of the house will have different parameters of sun and shade, and therefore heat and cold. To give you flexibility in choosing plants that fit the conditions in your yard, we have made the regional designations below for plants very broad. Please make sure to research appropriate choices for your specific area from the general plant lists provided.

A number of plant species have very wide distribution across the nation. Populations from different areas will be more or less adapted to different soils, heat, sunlight, shade, drought, humidity, and cold. That's why it is so important to buy plants as locally as possible. Always read the descriptions of conditions in which plants grow best at nurseries or in catalogs.

Further information can be obtained from books, local or mail order nurseries, Master Gardeners, University Extension Offices, and horticultural organizations. Visiting botanical gardens and public gardens, as well as attending educational events at these gardens, is an excellent way to gain information about gardening locally.

Hardiness Zone Number

The hardiness zone numbers in the Regional Plant Lists denote the low range of temperatures that the plants can tolerate. See the chart below.

HARDINESS ZONE NUMBER	LOWEST TEMPERATURE PLANT CAN TOLERATE
1	-60°F to -50°F
2	-50°F to -40°F
3	-40°F to -30°F
4	-30°F to -20°F
5	-20°F to -10°F
6	-10°F to 0°F
7	0°F to 10°F
8	10°F to 20°F
9	20°F to 30°F
10	30°F to 40°F
11	40°F to 50°F
12	50°F to 60°F
13	60°F to 70°F

All plants with a single hardiness zone number listed in the following charts will be hardy to the lowest temperature listed above. If a plant has a range of numbers listed, it means that the plant needs winter chill in order to flower. The plant may grow in areas with higher annual temperatures, but it may not bloom. For example, a plant that is listed as 3–8 may grow well in zone 10, but won't flower well there.

Even a very small difference in elevation can determine whether a plant freezes or doesn't as cold air flows downhill and tends to pool at the lowest area. Duration of cold is an important factor in plant hardiness as well. For instance, if you live in the lower areas of a valley, you may find that a certain plant doesn't survive the winter because cold air pools there, while the same plant may survive in your neighbors' yard because it's slightly higher with better cold air drainage.

If a plant is marked "summer annual," it means that it's an annual plant that only grows in frost-free times of the year. It must be planted when all danger of frost has passed. If a plant is marked "summer annual" but also has a number, it means that it can tolerate some cold temperatures as indicated by the hardiness zone number, but not a lot. It can be planted earlier in spring than the pure summer annuals, which cannot tolerate any frost.

Native Plants

Native plants are marked with an "N." See page 76 for a discussion of the benefits of native plants.

Invasive Plants

When a nonnative plant is marked "I" for invasive on the Regional Plant Lists, it can spread out of control in some ecosystems and climates. Caution is urged in use of these plants.

A federal list of invasive plants can be found at www.invasivespeciesinfo.gov/plants/main.shtml. Each state also may have more locally invasive plants. Information can be obtained online, at University Extension Offices, and from Master Gardeners organizations. We do not recommend planting any of these plants.

Spreading Plants

Some plants, both native and nonnative, tend to spread in gardens. They can be marvelous garden specimens, but care should be exercised in determining where to place them. They can spread in gardens if conditions are right, which may or may not be a desirable trait depending on the garden and gardener. For example, some goldenrods, some asters, and some milkweeds can spread over a large area. This could be very useful if you want to cover a large area of space with a ground cover, or if you are planting in a space with well-defined borders that will control the plant, like a container. These plants could do well in a boulevard (the space between the sidewalk and the street), which is bordered by concrete. They are often tough plants, and can tolerate foot traffic. When a plant is marked "S," it means it can potentially spread aggressively in a garden situation.

SOUTHEAST REGION

Alabama, Florida, Georgia, North Carolina, South Carolina, Tennessee

Horticultural Name	Common Name	Hardiness Zone Number	Native(N)/Invasive(I)/ Spreading(S)
ANNUALS			
Coriandrum sativum	cilantro	summer annual	
Cosmos bipinnatus	cosmos	summer annual	
Cosmos sulphureus	Klondike cosmos	summer annual	
Cuphea × 'Kirsten's Delight'	cigar plant	8	
Echium plantaginoides 'Blue Bedder'	'Blue Bedder' viper's bugloss	8	I
Gaillardia pulchella	annual blanket flower	4	N
Helianthus annus	annual sunflower	summer annual 9	N
Ocimum basilicum	basil	summer annual	
Rudbeckia hirta	black-eyed susan	9	N
Salvia coccinea	scarlet sage	9	
Tithonia rotundifolia	Mexican sunflower	summer annual	
PERENNIALS			
Asclepias tuberosa	butterfly weed	3–9	N
Aster carolinianus	climbing Carolina aster	6–9	N
Bidens ferulifolia	bidens	9	N
Bulbine frutescens	bulbine	8	
Dalea pinnata	summer farewell	6	N
Cassia fasciculate	partridge pea	5	N
Coreopsis grandiflora	large-flower tickseed	5	N
Echinacea purpurea	purple coneflower	2	N
Erigeron quercifolius	oakleaf fleabane	6	N
Eutrochium maculatum	joe-pye weed	4–5	N
Geranium maculatum	spotted geranium	4	N
Helenium autumnale	common sneezeweed	3	N
Helianthus debilis	beach sunflower	8	N
Liatris spicata	dense blazing star	3	N
Monarda punctata	spotted bee balm	5	N
Ocimum kilimandscharicum basilicum 'Dark Opal'	African blue basil	9	
Penstemon laevigatus	Eastern smooth penstemon	7	N
Penstemon multiflorus	many-flowered penstemon	8	N
Pentas lanceolata	pentas	9	
Phyla nodiflora	turkey tanglefoot	6	N
Pycnanthemum muticum	mountain mint	3–9	N

Horticultural Name	Common Name	Hardiness Zone Number	Native(N)/Invasive(I)/ Spreading(S)
Rudbeckia fulgida	black-eyed susan	5	N
Salvia azurea	azure sage	5	N
Salvia darcyi	Galeana red sage	7	
Salvia farinacea	mealycup sage	8	N
Salvia gauranitica	anise leaf sage	7	S
Salvia gregii	autumn sage	6	N
Salvia 'Indigo Spires'	'Indigo Spires' sage	7	
Salvia 'Mystic Spires'	'Mystic Spires' sage	8	
Sedum telephium 'Autumn Joy'	'Autumn Joy' sedum	7	
Stokesia laevis	Stokes' aster	5–9	N
Tagetes lemonii	Mexican marigold	9	
Teucrium canadense	American teucrium	5	N
Tradescantia virginiana	Virginia spiderwort	5	N
Vernonia gigantea	giant ironweed	5	N
SHRUBS			
Abelia grandiflora	glossy abelia	6	
Callistemon citrinus	red bottlebrush	8	
Cercis canadensis	Eastern redbud	4–9	N
Citharexylum fruticosum	fiddlewood	10	N
Ilex glabra	inkberry	5–9	N
Lyonia ferruginea	rusty lyonia	8	N
Passiflora incarnata	purple passion flower	5	N
Persea borbonia	red bay	7	N
Physocarpus opulifolius	common ninebark	2–9	N
Rosa carolina	Carolina rose	5	N
Ruellia caroliniensis	wild petunia	8	N
Serenoa repens	saw palmetto	7	N
Vaccinium arboreum	sparkleberry	6–9	N
Vaccinium corymbosum	highbush blueberry	3–8	N
Vaccinium viagatum	rabbiteye blueberry	7–10	N
Viburnum obovatum	Walter's viburnum	7	N
Vitex agnus-castus	chaste tree	6	
TREES			
Acer rubrum	red maple	3–9	N
Ilex attenuata	Eastern Palatka holly	7–9	N

SOUTHEAST REGION, CONTINUED

Horticultural Name	Common Name	Hardiness Zone Number	Native(N)/Invasive(I)/ Spreading(S)
Ilex opeca	American holly	5	N
Liriodendron tulipfera	tulip tree	5–9	N
Magnolia grandiflora	Southern magnolia	5–9	N
Nyssa ogeche	Ogeechee tupelo	7–9	N
Oxydendron arboreum	sourwood tree	5–9	N
Prunus caroliniana	Carolina laurel cherry	7	N
Sabal palmetto	cabbage palm	8	N

SOUTH CENTRAL REGION

Arkansas, Louisiana, Mississippi, Oklahoma, Texas

Horticultural Name	Common Name	Hardiness Zone Number	Native(N)/Invasive(I)/ Spreading(S)
ANNUALS			
Cleome hasslerana	spider flower	summer annual	
Coreopsis tinctoria	plains coreopsis	summer annual	
Coriandrum sativum	cilantro	summer annual	
Cosmos bipinnatus	cosmos	summer annual	
Cosmos sulphureus	Klondike cosmos	summer annual	
Cuphea spp.	cigar plant	9	N
Echium plantagineum	viper's bugloss	8	I
Gaillardia pulchella	annual blanket flower	4	N
Helianthus annuus	annual sunflower	summer annual	N
Ocimum basilicum	basil	summer annual	
Papaver rhoes	Shirley poppy	8	
Phacelia tanacetifolia	lacy phacelia	8	N
Rudbeckia hirta	black-eyed susan	9	N
Tithonia rotundifolia	Mexican sunflower	summer annual	
PERENNIALS			
Agastache spp.	hummingbird mint	varies by species	
Agastache 'Blue Blazes'	'Blue Blazes' hummingbird mint	5	N
Agastache 'Purple Haze'	'Purple Haze' hyssop	6	N
Agastache × foeniculum hybrids	licorice mint	5	N
Asclepias tuberosa	butterfly weed	3	N
Asters spp.	asters	5	
Baptisia alba	wild white indigo	3–9	N
Coreopsis grandiflora	large-flower tickseed	5	N
Cuphea spp.	cigar plant	9	

Horticultural Name	Common Name	Hardiness Zone Number	Native(N)/Invasive(I)/ Spreading(S)
Dalea purpurea	purple prairie clover	3	N
Dracopis amplexicaulis	clasping coneflower	5	N
Echinacea purpurea	purple coneflower	2	N
Eryngium yuccifolium	rattlesnake master	5	N
Eutrochium maculatum	joe-pye weed	5	N
Helenium autumnale	sneezeweed	3	N
Liatris spp.	prairie blazing star	3	
Monarda fistulosa	bergamot	4	N
Nepata × faasseni	catmint	4	
Origanum spp.	oregano	varies by species	
Penstemon spp.	penstemon	3–10	N
Perovskia atriplicifolia	Russian sage	4	
Phlox pilosa	prairie phlox	5–8	N
Pycnanthemum muticum	mountain mint	3–9	N
Rosmarinus officinalis	rosemary	varies by cultivar	
Scabiosa spp.	pincushion flower	varies by species	
Scabiosa caucasica 'Fama Blue'	'Fama Blue' pincushion flower	4	
Sedum telephium 'Autumn Fire'	'Autumn Fire' sedum	4	
Solidago spp.	goldenrod	5	N, S
Tradescantia virginiana	Virginia spiderwort	5	N
SHRUBS			
Abelia grandiflora	glossy abelia	6	N
Amorpha canescens	leadplant	2–9	N
Amorpha fruticosa	wild indigo	4–9	N
Callicarpa americana	beautyberry	6	N
Ceanothus americanus	New Jersey tea	4–8	N
Cercis canadensis	Eastern redbud	4–8	N
Clethra alniflora	common sweet pepperbush	4	N
Cliftonia monophylla	spring titi	7	N
Ilex glabra	inkberry	5–9	N
Itea virginica	Virginia sweetspire	5	N
Lagerstromeria indica × fauriei 'Natchez'	'Natchez' crape myrtle	6	
Rhus copallina	winged sumac	5	N
Rosa carolina	Carolina rose	4–9	N
Sabal palmetto	cabbage palmetto	8	N
Vaccinium corymbosum	highbush blueberry	3	N
Viburnum nudum	possumhaw	5	N

SOUTH CENTRAL REGION, CONTINUED

Horticultural Name	Common Name	Hardiness Zone Number	Native(N)/Invasive(I)/Spreading(S)
Viburnum rufidulum	rusty blackhaw	5–9	N
Vitex agnus-castus	chaste tree	6	
TREES			
Acer rubrum	red maple	3–9	N
Aesculus pavia	red buckeye	4–9	N
Cartaegus crus-galli	cockspur hawthorn	4	N
Catalpa bignonioides	southern catalpa	5	N
Crataegus spp.	hawthorn	5	
Ilex glabra	inkberry	5–9	N
Ilex opeca	American holly	5	N
Liriodendron tulipifera	tulip tree	5	N
Magnolia grandiflora	southern magnolia	4–9	N
Magnolia virginiana	sweet bay	5	N
Malus	flowering crab apple	5	N
Nyssa ogeche	Ogeechee tupelo	7–9	N
Nyssa sylvatica	tupelo	5	N
Prunus caroliniana	Carolina laurel cherry	7	N
Robinia pseudoacacia	black locust	4	N
Styrax americanus	American snowbell	5	N
Styrax grandiflorus	bigleaf snowbell	7	N

SOUTHWEST REGION

Arizona, California (Southern), Nevada, New Mexico, Utah

Horticultural Name	Common Name	Hardiness Zone Number	Native(N)/Invasive(I)/Spreading(S)
ANNUALS			
Argemone platycerus	busy bee prickly poppy	8	N
Argemone pleicantha	white prickly poppy	3	N
Coreopsis tinctoria	plains coreopsis	summer annual 9	N
Cosmos sulphureus	klondyke cosmos summer annual		
Eschscholtzia californica ssp. *mexicana*	Mexican gold poppy	7	N
Gaillardia pulchella	blanket flower	9	N
Helianthus annuus	annual sunflower	9	N
Kallstroemia grandiflora	Arizona poppy	7	
Lupinus arizonicus	Arizona lupine	8	
Ocimum basilicum	basil	summer annual	

Horticultural Name	Common Name	Hardiness Zone Number	Native(N)/Invasive(I)/ Spreading(S)
Phacelia spp.		8	N
Phacelia campanularia	desert bluebells	8	N
Phacelia minor	wild Canterbury bells	8	N
Phacelia tanacetifolia	lacy phacelia	8	N
Salvia columbariae	chia	8	N
PERENNIALS			
Agastache spp.	hummingbird mint	varies by species	N
Agastache 'Blue Blazes'	'Blue Blazes' hummingbird mint	5	N
Agastache cana	Texas hummingbird mint	3	N
Agastache neomexicana	New Mexico hummingbird mint	5	N
Agastache 'Purple Haze'	'Purple Haze' hyssop	6	N
Baileya multiradiata	desert marigold	6	N
Dalea purpurea	purple prairie clover	3	N
Eriogonum spp.	wild buckwheat	varies by species	N
Eriogonum umbellatum	sulphur-flower buckwheat	4	N
Gaillardia aristata	blanket flower	4	N
Gaillardia × grandiflora	blanket flower	5	N
Gaura lindheimeri	gaura	6	N
Glandularia tenuisecta	moss verbena	8	N
Helianthus maximilianii	New Mexico sunflower	2	N
Liatris punctata	prairie blazing star	3	N
Penstemon grandiflorus	large-flowered penstemon	3	N
Penstemon mexicali 'Pike's Peak Purple'	'Pike's Peak Purple' hybrid penstemon	5–9	N
Penstemon palmeri	Palmer's penstemon	4	N
Penstemon parryi	Parry's penstemon	8	N
Penstemon pseudospectabilis	desert penstemon	5	N
Penstemon superbus	large-flowered penstemon	7	N
Rudbeckia hirta	black-eyed susan	4	N
Salvia arizonica	deep-blue Arizona sage	5	N
Salvia dorii	desert purple sage	5	N
Salvia farinacea	mealy cup sage	8	N
Sphaeralcea ambigua	globe mallow	4	N
Sphaeralcea coccinia	scarlet globe mallow	5	N
Sphaeralcea monroana	Monro's globe mallow	4	N
Thymophylla pentachaeta	golden fleece	7	N
Verbesina encelioides	golden crownbeard	8	N

SOUTHWEST REGION, CONTINUED

Horticultural Name	Common Name	Hardiness Zone Number	Native(N)/Invasive(I)/ Spreading(S)
SHRUBS			
Acacia farenesiana	sweet acacia	7	N
Acacia wrightii	Wright acacia	6	N
Aloysia gratissima	bee bush	8	N
*Arctostaphylos pungens**	pointleaf manzanita	7	N
Baccharis glutinosa	mule fat	7	N
Calliandra californica	Baja fairy duster	9	N
Calliandra eriophylla	fairy duster	7	N
Callistemon citrinus	red bottlebrush	8	N
*Ceanothus fendleri**	Fendler's ceanothus	5	N
Ceanothus greggii	desert ceanothus	5	N
Cercis canadensis var. *mexicana*	Mexican redbud	5–9	N
Chrysothamnus nauseosus	yellow twig rabbitbrush	4	N
Dalea spp.	dalea	varies by species	N
Dalea bicolor	silver dalea	9	N
Dalea frutescens	broom dalea	6	N
Dalea pulchra	indigo bush	7–8	N
Dasylirion wheeleri	sotol	7	N
Encelia farinosa	brittlebush	8	N
Fallugia paradoxa	apache plume	6	N
Larrea tridentata	creosote bush	7	N
Leucophyllum spp.	Texas ranger	7	N
Leucopyllum langmaniae	Texas ranger	7	N
Nolina bigelovii	Bigelow's nolina	9	N
Rhus microphylla	littleleaf sumac	6	N
Rhus ovata	sugar bush	8	N
Salvia clevelandii	Cleveland sage	7	N
Salvia greggii	autumn sage	7	N
Salvia × 'Celestial Blue'	'Celestial Blue' sage	6	N
Salvia × 'Pozo Blue'	'Pozo Blue' sage	7	N
Tecoma stans	yellow bells	7	N
Vitex agnus-castus	chaste tree	6	N
***Best for high elevations**			
TREES			
Acacia constricta	whitethorn acacia	6	N
Acacia gregii	catclaw acacia	7	N

Horticultural Name	Common Name	Hardiness Zone Number	Native(N)/Invasive(I)/ Spreading(S)
Chilopsis linearis	desert willow	6	N
Parkinsonia (Cercidium) floridum	blue palo verde	7	N
Parkinsonia (Cercidium) microphylla	palo verde cultivars and hybrids	8	N
Parkinsonia (Cercidium) × 'Desert Museum'	'Desert Museum' palo verde	8	N
Prosopis spp.	mesquite	varies by species	
Prosopis glandulosa	Texas honey mesquite	6	N
Prosopis velutina	velvet mesquite	7	N
Sophora secundiflora	mescal bean	7	N
CACTI			
Agave parryi huachucensis	huachuca agave	7	N
Carnegiea gigantea	saguaro	9	N
Fouquieria splendens	ocotillo	8	N
Opuntia spp.	prickly pear	varies by species and cultivars 3–11	N

PACIFIC NORTHWEST REGION

California (Northern), Idaho, Oregon, Washington

Note: Many of the plants for the Northeast/Mid-Atlantic will grow in the Pacific Northwest.

Horticultural Name	Common Name	Hardiness Zone Number	Native(N)/Invasive(I)/ Spreading(S)
ANNUALS			
Anchusa capansis	forget-me-not	8	
Argemone platycerus	prickly poppy	8	
Borago officinalis	borage	8	
Centaurea cyanus	bachelor button	8	
Cerinthe major 'Purpurascens'	honeywort	summer annual 8	
Cleome hasslerana	spider flower	summer annual	
Collinsia heterophylla	Chinese houses	8	N
Coreopsis tinctoria	plains coreopsis	summer annual	
Cosmos bipinnatus	cosmos	summer annual	
Cosmos sulphureus	sulphur cosmos	summer annual	
Cuphea × 'Kirsten's Delight'	cigar plant	8	
Echium plantagineum 'Blue Bedder'	'Blue Bedder' viper's bugloss	8	I
Eschscholzia californica	California poppy	8	N
Helianthus annuus	annual sunflower	summer annual	N

PACIFIC NORTHWEST REGION, CONTINUED

Horticultural Name	Common Name	Hardiness Zone Number	Native(N)/Invasive(I)/ Spreading(S)
Limnanthes douglasii	meadowfoam	8	N
Nemophila menziesii	baby blue eyes	8	N
Papaver hybridum	peony poppy	8	
Papaver rhoes	Shirley poppy	8	
Phacelia bolanderi	Bolander's phacelia	8	N
Phacelia campanularia	desert bluebells	8	N
Phacelia minor	wild canterbury bells	8	N
Phacelia tanacetifolia	lacy phacelia	8	N
Rudbeckia trilobata	brown-eyed susan	4	N
Scabiosa atropurpurea	pincushion flower	8	
PERENNIALS			
Agastache spp.	hummingbird mint	varies by species	
Agastache 'Blue Blazes'	'Blue Blazes' hummingbird mint	5	N
Agastache 'Purple Haze'	'Purple Haze' hyssop	6	N
Angelica stricta 'Purpurea'	purple angelica	5	
Agastache × foeniculum hybrids	licorice mint	5	N
Aquilegia spp.	columbine	3	N
Asclepias speciosa	showy milkweed	4	N
Aster (Symphytrichum) ericoides 'Monte Cassino'	heath aster	5	N
Aster × frikarti 'Monch'	Frikart's aster	5	
Bidens ferulifolia	bidens	9	N
Caryopteris incana	bluebeard	5	
Clinopodium nepeta	calamint	5–10	
Coreopsis grandiflora	large-flower tickseed	5	N
Digitalis purpurea	foxglove	4	
Echinops bannaticus	globe thistle	3	
Erigeron glaucus 'Bountiful'	'Bountiful' seaside daisy	3	
Erigeron glaucus 'Wayne Roderick'	'Wayne Roderick' seaside daisy	3	
Eriogonum spp.	wild buckwheat	8	N
Eriogonum fascicularis	California buckwheat	6	N
Eryngium × tripartitum	blue sea holly	5	N
Erysimum 'Bowle's Mauve'	'Bowle's Mauve' wallflower	6	
Gaillardia × grandiflora 'Oranges and Lemons'	'Oranges and Lemons' blanket flower	3	N
Geranium spp.	hardy geranium	varies by species and cultivars	
Grindelia camporum	valley gumweed	8	N

Horticultural Name	Common Name	Hardiness Zone Number	Native(N)/Invasive(I)/ Spreading(S)
Helenium autumnale hybrids	common sneezeweed	3	N
Lavandula spp.	lavender	5	
Maianthemum racemosum	false Solomon's seal	3–8	N
Marrubium supinum	perennial horehound	8	
Monarda fistulosa	bergamot	4	
Monardella villosa	coyote mint	8	N
Nepeta × faassenii	catmint	4	
Origanum 'Betty Rollins'	'Betty Rollins' oregano	5	
Origanum 'Bristol Cross'	'Bristol Cross' oregano	5	
Origanum 'Santa Cruz'	'Santa Cruz' oregano	5	
Penstemon spp.	penstemon	varies by species	
Penstemon heterophyllus	foothill penstemon	6	N
Perovskia atriplicifolia	Russian sage	4	
Persicaria amplexicaulis 'Taurus'	'Taurus' knotweed	5	
Salvia 'Bee's Bliss'	'Bee's Bliss' sage	8	N
Salvia brandegeei 'Pacific Blue'	'Pacific Blue' Brandegee's sage	8	N
Salvia clevelandii × leucophlla 'Pozo Blue'	'Pozo Blue' sage	7	N
Salvia forsskaolii	woodland indigo sage	5	
Salvia gauranitica	anise leaf sage	7	S
Salvia gregii	autumn sage	7	N
Salvia mellifera	black sage	8	N
Salvia sclarea	clary sage	5	
Salvia uliginosa	bog sage	6	I
Salvia × 'Celestial Blue'	'Celestial Blue' sage	7	N
Salvia × 'Indigo Spires'	'Indigo Spires' sage	8	
Salvia × 'Mystic Spires'	'Mystic Spires' blue sage	8	
Salvia × 'Purple Majesty'	'Purple Majesty' sage	8	
Scabiosa caucasica 'Fama Blue'	'Fama Blue' pincushion flower	4	
Solidago spp.	goldenrod	varies by species	N, S
Solidago californica	California goldenrod	6	N, I
Sphaeralcea coccinea	scarlet globe mallow	6	N
Sphaeralcea munroana	Monro's globe mallow	4	N
Sedum telephium 'Autumn Fire'	'Autumn Fire' sedum	4	
Teucrium cussoni 'Majoricum'	fruity teucrium	8	
Teucrium chamaedrys	wall germander	5	
Teucrium hircanicum	wood sage	5	

PACIFIC NORTHWEST REGION, CONTINUED

Horticultural Name	Common Name	Hardiness Zone Number	Native(N)/Invasive(I)/ Spreading(S)
Thymus spp.	creeping thyme, bush thyme	5	
Verbascum olympicum	Greek mullien	5	
Verbena bonariensis	tall verbena	7	
SHRUBS			
Arbutus unedo	strawberry tree	7	
Arctostaphylos spp.	manzanita	varies by species	N
Arctostaphylos bakeri 'Louis Edmunds'	'Louis Edmund' manzanita	7	N
Arctostaphylos densiflora 'Howard McMinn'	'Howard McMinn' manzanita	7	N
Berberis spp.	mahonia	varies by species	N
Ceanothus spp.	California lilac	varies by region	
Ceanothus 'Joyce Coulter'	'Joyce Coulter' California lilac	8	N
Ceanothus 'Julia Phelps'	'Julia Phelps' California lilac	8	N
Ceanothus 'Ray Hartman'	'Ray Hartman' California lilac	8	N
Ceanothus 'Skylark'	'Skylark' California lilac	8	N
Ceanothus maritimus 'Valley Violet'	'Valley Violet' California lilac	8	N
Cercis occidentalis	California redbud	6–8	N
Cotoneaster dammeri	bearberry cotoneaster	5–9	
Crataegus douglasii	black hawthorn	5	N
Crataegus spp.	hawthorn	4	
Escallonia spp.	escallonia	8	
Escallonia rubra	escallonia	8–10	
Heteromeles arbutifolia	toyon	8	N
Holodiscus discolor	oceanspray	6–9	N
Lavatera thuringiaca	tree mallow	7	N
Myrtus communis 'Compacta'	compact myrtle	8	
Physocarpus opulifolius	common ninebark	2–9	N
Rhamnus californica	coffeeberry	7	N
Rhus ovata	sugar bush	8	N
Rosmarinus officinalis	rosemary	6	N
Spirea douglasii	Douglas spirea	5	N
Symphoricarpus albus	common snowberry	6	N
Vitex agnus-castus	chaste tree	6	
TREES			
Acer macrophyllum	Western bigleaf maple	4	N
Arbutus menziesii	madrone	7	N

Horticultural Name	Common Name	Hardiness Zone Number	Native(N)/Invasive(I)/ Spreading(S)
Arbutus 'Marina'	'Marina' strawberry tree	8	N
Crataegus spp.	hawthorn	varies by species 4–9	N
Lagerstroemia indica × fauriei 'Natchez'	crape myrtle	6	
Liriodendron tulipifera	tulip tree	5–9	N
Magnolia grandiflora	southern magnolia	4–9	N
Magnolia virginiana	sweet bay	5	N
Malus spp.	flowering crab apple	4–9	
Nyssa sylvatica	tupelo	5	N
Prunus spp.	plum	4	
Rhus typhinia	staghorn sumac	3–8	N
Sapium sebiferum	Chinese tallow tree	8	I
Styrax japonica	Japanese snowbell tree	6	

ROCKY MOUNTAIN/INTERMOUNTAIN WEST REGION

Colorado, Kansas, Montana, Nebraska, North Dakota, South Dakota, Wyoming

Horticultural Name	Common Name	Hardiness Zone Number	Native(N)/Invasive(I)/ Spreading(S)
ANNUALS			
Anchusa capansis	forget-me-not	8	
Argemone platyceras	prickly poppy	summer annual	N
Borago officinalis	borage	8	
Centaurea cyanus	bachelor button	8	
Cerinthe major 'Purpurascens'	honeywort	8	
Cleome hasslerana	spider flower	summer annual	
Cleome lutea	yellow bee flower	summer annual	N
Cleome serrulata	Rocky Mountain bee plant	summer annual	N
Coreopsis tinctoria	plains coreopsis		N
Cosmos bipinnatus	cosmos	summer annual	
Cosmos sulphureus	Klondike cosmos	summer annual	
Echium plantagineum 'Blue Bedder'	'Blue Bedder' viper's bugloss	8	
Helianthus annuus	annual sunflower	summer annual	N
Papaver hybridum	peony poppy	8	
Papaver rhoes	Shirley poppy	8	
Phacelia campanularia	desert bluebells	8	N
Phacelia tanacetifolia	lacy phacelia	8	N
Rudbeckia hirta	black-eyed susan	summer annual	N

ROCKY MOUNTAIN/INTERMOUNTAIN WEST REGION, CONTINUED

Horticultural Name	Common Name	Hardiness Zone Number	Native(N)/Invasive(I)/ Spreading(S)
Rudbeckia trilobata	brown-eyed susan	4	N
Scabiosa atropurpurea	pincushion flower	summer annual 8	
PERENNIALS			
Asclepias speciosa	showy milkweed	4	N
Aster novae-angliae	New England aster	3	N
Aster novae-belgii	New York aster	4	N
Agastache spp.	hummingbird mint	varies by species	N
Agastache 'Blue Blazes'	'Blue Blazes' hyssop	5	N
Agastache 'Purple Haze'	'Purple Haze' hyssop	6	N
Asclepias tuberosa	butterfly weed	3	N
Campanula rotundifolia	bellflower	3	N
Dalea purpurea	purple prairie clover	3	N
Echinacea purpurea	purple coneflower	2	N
Echinocereus viridiflorus	prairie hedgehog cactus	3	N
Echinops bannaticus	blue globe thistle	3	
Eriogonum niveum	snow buckwheat	4	N
Eriogonum umbellatum polyanthemum	sulphur-flower buckwheat	3	N
Eryngium spp.	blue sea holly	varies by species	
Eryngium planum	blue sea holly	4	N
Gaillardia aristata	blanket flower	3	N
Gaillardia × *grandiflora*	blanket flower	3	N
Helianthus maximilianii	Maximilian's sunflower	4	N
Lavandula angustifolia	English lavender	5	
Liatris punctata	prairie blazing star	3	N
Monarda fistulosa	bergamot	4	N
Nepeta faassenii	catmint	4	
Oreganum 'Amethyst Falls'	'Amethyst Falls'oregano	5	
Penstemon digitalis	foxglove penstemon	3	N
Penstemon mexicali	penstemon	5	N
Penstemon palmeri	Palmer's penstemon	4	N
Penstemon pseudospectabilis	desert penstemon	5	N
Penstemon strictus	Rocky Mountain penstemon	2	N
Perovskia atriplicifolia	Russian sage	4	
Salvia arizonica 'Deep Blue'	'Deep Blue' Arizona sage	5–9	N
Salvia azurea	prairie sage	4	N
Salvia gregii 'Furman's Red'	'Furman's Red' autumn sage	6	

Horticultural Name	Common Name	Hardiness Zone Number	Native(N)/Invasive(I)/ Spreading(S)
Salvia nemorosa (sylvestris)	woodland sage	3	
Salvia pachyphylla 'Blue Flame'	'Blue Flame' purple sage	5–9	N
Salvia sclarea	clary sage	5	
Salvia verticilata 'Purple Rain' or 'Endless Love'	lilac sage	5	N
Scabiosa caucasica 'Fama Blue'	'Fama Blue' pincushion flower	4	
Sedum telephium 'Autumn Fire'	'Autumn Fire' sedum	4	
Solidago 'Wichita Mountains'	'Wichita Mountains'goldenrod	4	N, S
Solidago sphacelata	dwarf goldenrod	4	N, S
Sphaeralcea coccinea	scarlet globe mallow	6	N
Sphaeralcea monroana	Monro's globe mallow	4	N
Symphyotrichum ericoides	heath aster	3	N
Symphyotrichum oblongifolium	aromatic aster	3	N
Symphyotrichum porteri	smooth white aster	3	N
Teucrium cussonii	gray germander	5	
Thymus spp.	thyme, creeping thyme	4	
Vernonia lindheimeri	silver ironweed	4	N
SHRUBS			
Amelanchier alnifolia	serviceberry	2–9	N
Amelanchier utahensis	Utah serviceberry	3–9	N
Arctostaphylos pungens	pointleaf manzanita	5	N
Berberis fendleri	Fendler's barberry	4	N
Chrysothamnus nauseosus	yellow twig rabbitbrush	4	N
Cotoneaster intergerrimus	cotoneaster	3	N
Crataegus spp.	hawthorn	4	
Ericameria nauseosa var. *hololeuca* 'Santa Fe Silver'	rubber rabbitbrush	4	N
Fallugia paradoxa	apache plume	6	N
Mahonia fremontii	Fremont's mahonia	5–9	N
Mahonia haematocarpa	red barberry	7	N
Nolina microcarpa	sacahuista	5	N
Opuntia phaeacantha	tulip prickly pear	4	N
Philadelphus lewesii	Western mock orange	4	N
Philadelphus microphyllus	littleleaf mock orange	3	N
Physocarpus opulifolius	common ninebark	2–9	N
Prunus besseyi	sand cherry	3–9	
Prunus tomentosa	Nanking cherry	2–9	

ROCKY MOUNTAIN/INTERMOUNTAIN WEST REGION, CONTINUED

Horticultural Name	Common Name	Hardiness Zone Number	Native(N)/Invasive(I)/ Spreading(S)
Rhus trilobata	three-leaf sumac	4	N
Ribes aureum	golden currant	3	N
TREES			
Acer glabrum	Rocky Mountain maple	5	N
Crataegus spp.	hawthorn	5	
Crataegus douglasii	black hawthorn	5	N
Malus	flowering crab apples	4–8	N
Prunus virginiana	chokecherry	3	N

NORTHEAST/MIDWEST/MID-ATLANTIC REGION

Connecticut, Delaware, Illinois, Indiana, Iowa, Kentucky, Maine, Maryland, Massachusetts, Michigan, Minnesota, Missouri, New Hampshire, New Jersey, New York, Ohio, Pennsylvania, Rhode Island, Vermont, Virginia, Washington DC, West Virginia, Wisconsin

Horticultural Name	Common Name	Hardiness Zone Number	Native(N)/Invasive(I)/ Spreading(S)
ANNUAL			
Anchusa capansis	forget-me-not	summer annual 8	
Borago officinalis	borage	summer annual 8	
Centaurea cyanus	bachelor button	summer annual 8	
Cerinthe major 'Purpurascens'	honeywort	summer annual	
Cleome hasslerana	spider flower	summer annual	
Coreopsis tinctoria	plains coreopsis	summer annual	N
Cosmos bipinnatus	cosmos	summer annual	
Cosmos sulphureus	Klondike cosmos	summer annual	
Echium plantagineum 'Blue Bedder'	'Blue Bedder' viper's bugloss	summer annual	I
Helianthus annuus	annual sunflower	summer annual	N
Limnanthes douglasii	meadowfoam	summer annual	
Papaver hybridum	peony poppy	summer annual	
Papaver rhoes	Shirley poppy	summer annual 8	
Phacelia tanacetifolia	lacy phacelia	summer annual	
Rudbeckia hirta	black-eyed susan	summer annual	N
Rudbeckia trilobata	brown-eyed susan	6	N
Scabiosa atropurpurea	pincushion flower	summer annual	
PERENNIALS			
Agastache 'Blue Fortune'	'Blue Fortune' hybrid hyssop	4	N
Agastache 'Purple Haze'	'Purple Haze' hyssop	6	N
Anemone canadensis	Canada anemone	3	N
Aquilegia canadensis	eastern red columbine	3–8	N

Horticultural Name	Common Name	Hardiness Zone Number	Native(N)/Invasive(I)/ Spreading(S)
Asclepias syriaca	common milkweed	4	N
Asclepias tuberosa	butterfly weed	3	N
Aster (Symphytrichum) ericoides 'Monte Cassino'	heath aster	5	N
Aster novae-angliae	New England aster	3	N
Aster novae-belgii	New York aster	4	N
Campanula rotundifolia	bellflower	3	N
Clinopodium nepeta	calamint	5–10	
Dalea purpurea	purple prairie clover	3	N
Echinacea purpurea	purple coneflower	2	N
Echinops bannaticus	globe thistle	3	
Eryngium yuccifolium	rattlesnake master	5	N
Eupatorium perfoliatum	common boneset	4	N, S
Geranium spp.	hardy geranium	varies by species and cultivars	
Geranium macrorrhizum	scented cranesbill	3	
Geranium maculatum	spotted geranium	4	N
Helenium spp. and hybrids	sneezeweed	3	N
Helianthus maximilianii	Maximilian's sunflower	4	N
Heliopsis helianthoides	false sunflower	3	N
Lavandula angustifolia	English lavender	5	N
Liatris spicata	prairie blazing star	3–9	N
Lupinus perennis	wild lupine	3	N
Maianthemum racemosum	false solomon's seal	4	N
Monarda fistulosa	bergamot	4	N
Nepeta fassenii 'Walker's Low'	catmint	4	
Pentstemon digitalis	foxglove penstemon	3	N
Perovskia atriplicifolia	Russian sage	4	N
Pycnanthemum muticum	mountain mint	3–9	N
Ratibida pinnata	yellow coneflower	3–8	N
Rudbeckia fulgida 'Goldstrum'	'Goldstrum' coneflower	5	N
Scabiosa caucasica 'Fama Blue'	'Fama Blue' pincushion flower	4	
Solidago 'Wichita Mountains'	'Witchita Mountains' goldenrod	4	N, S
Solidago rugosa 'Fireworks'	'Fireworks' goldenrod	4	N, S
Solidago sphacelata 'Golden Fleece'	'Golden Fleece' dwarf goldenrod	4–9	N, S
Sylphium perfoliatum	cup plant	4–8	N
Verbena stricta	hoary vervain	3	N
Vernonia fasciculata	ironweed	3	N

NORTHEAST/MIDWEST/MID-ATLANTIC REGION, CONTINUED

Horticultural Name	Common Name	Hardiness Zone Number	Native(N)/Invasive(I)/ Spreading(S)
Vernonia lettermani	iron butterfly weed	4	N
Veronicastrum virginicum	Culver's root	3	N
Zizia aurea	golden Alexanders	4	N
SHRUBS			
Amelanchier canadensis	serviceberry	4–7	N
Amorpha canescens	leadplant	2–9	N
Arctostaphylos uva-ursi	kinnikinnick	5–10	N
Berberis spp.	mahonia	varies by species	N
Caryopteris × 'Longwood Blue'	blue mist bush	5	
Ceanothus americanus	New Jersey tea	4–8	N
Cercis canadensis	Eastern redbud	4–8	N
Clethra alnifolia	summersweet	4	N
Cotoneaster intergerrimus	cotoneaster	3	N
Enkianthus campanulatus	redvein enkianthus	5	
Hamamelis virginianus	American witchhazel	3–8	N
Ilex glabra	inkberry	5–9	N
Ilex verticillata	common winterberry	3–9	
Itea virginica	Virginia sweetspire	5	N
Philadelphus coronarious	mock orange	5	
Physocarpus opulifolius	common ninebark	2–9	N
Prunus caroliniana	Carolina laurel cherry	7	N
Prunus virginiana	chokecherry	2–9	N
Spirea alba	white meadowsweet	3	N
Symphoricarpus albus	snowberry	4	N
Vaccinium angustifolium	lowbush blueberry	3–7	N
TREES			
Crataegus spp. and cultivars	hawthorn	3–9	N, depending on species and cultivar
Ilex opaca	American holly	5	N
Liriodenrdon tulipifera	tulip tree	5–9	N
Malus	flowering crab apples	4–8	
Nyssa sylvatica	tupelo	5	N
Oxydendron arboreum	sourwood tree	5–9	N
Rhus typhinia	staghorn sumac	3–8	N
Robinia pseudoacacia	black locust	4–8	N
Sorbus americana	American mountain ash	2–5	N
Tilia americana	American linden or basswood	3	N

Photography Credits

All photography by Leslie Lindell, except as noted:

Pages iv, 28 (top), 64 (top row, left and right; bottom row, right), 166: Susan Damon

Pages vi, 36, 44, 47, 53, 56 (top row, center row left and right, bottom row, left), 58, 59, 68 (bottom), 69, 72 (top row right, center row middle), 75 (bottom row), 90 (top row right), 102-103, 106, 113, 116, 118, 119, 120, 128, 138, 140, 143, 145, 146, 149, 161, 164, 165: Kate Frey

Page ix: Paul Asper

Page 9: Whole Foods Market

Pages 10, 19, 169 (bottom): Janet Allen

Pages 16, 24-27, 28 (bottom), 30, 174, 179: Rollin Coville

Page 20 (illustration): Ashley Lima

Page 22: Ron McGinley

Page 31, 33 (top), 127 (bottom): Heather Holm

Pages 32 (top), 60: Kathy Keatley Garvey

Page 32 (bottom): Jason Graham

Page 33 (bottom): Cheryl Fimbrel

Page 34 (top): Gretchen LeBuhn

Page 34 (bottom): David Inouye

Page 127 (top): Eva Johansson

Page 129: Jason Graham

Page 162: Megan O'Donald

Page 167: Michael McDowell

Page 168: Deb Bosworth

Page 169 (top): Lauren Springer Ogden

Special thanks to the gardens and gardeners who generously allowed us to photograph their gardens for this book: Maile Arnold, Bruce and Linda Berlinger, Digging Dog Nursery, Filoli, Jennifer Kearney of J. Kearney Design, Lynmar Estate Winery, The Melissa Garden, Mendocino Botanical Garden, Stanford University, Stone Edge Farm, Charlotte Torgovitsky, and the United States Botanic Garden.

index

Published in the United States by Ten Speed Press, an imprint of the Crown Publishing Group, a division of Penguin Random House LLC, New York.
www.crownpublishing.com
www.tenspeed.com

Ten Speed Press and the Ten Speed Press colophon are registered trademarks of Penguin Random House LLC.

All photographs by Leslie Lindell with the exception of those noted on page 207.

Library of Congress Cataloging-in-Publication Data
Frey, Kate, 1960- author.
The bee-friendly garden / Kate Frey and Gretchen LeBuhn. — First edition.
pages cm
1. Gardening to attract wildlife. 2. Bees. 3. Honey plants. I. LeBuhn, Gretchen, 1961- author. II. Title.
QL59.F74 2016
595.79'9—dc23
2015025815

Trade Paperback ISBN: 978-1-60774-763-5
eBook ISBN: 978-1-60774-764-2

Printed in China

Design by Ashley Lima

11 10 9 8 7 6 5 4

First Edition